Septic Shock Methods and Protocols

METHODS IN MOLECULAR MEDICINE™

John M. Walker, Series Editor

46. **Angiogenesis: *Reviews and Protocols*, edited by** *J. Clifford Murray*, 2000
45. **Hepatocellular Carcinoma Methods and Protocols, edited by** *Nagy A. Habib*, 2000
44. **Asthma: *Mechanisms and Protocols*, edited by** *K. Fan Chung and Ian Adcock*, 2000
43. **Muscular Dystrophy: *Methods and Protocols*, edited by** *Katherine B. Bushby and Louise Anderson*, 2000
42. **Vaccine Adjuvants: *Preparation Methods and Research Protocols*,** edited by *Derek T. O'Hagan*, 2000
41. **Celiac Disease: *Methods and Protocols*,** edited by *Michael N. Marsh*, 2000
40. **Diagnostic and Therapeutic Antibodies,** edited by *Andrew J. T. George and Catherine E. Urch*, 2000
39. **Ovarian Cancer: *Methods and Protocols*,** edited by *John M. S. Bartlett*, 2000
38. **Aging Methods and Protocols,** edited by *Yvonne A. Barnett and Christopher P. Barnett*, 2000
37. **Electrically Mediated Delivery of Molecules to Cells,** edited by *Mark J. Jaroszeski, Richard Heller, and Richard Gilbert*, 2000
36. **Septic Shock Methods and Protocols,** edited by *Thomas J. Evans*, 2000
35. **Gene Therapy of Cancer: *Methods and Protocols*,** edited by *Wolfgang Walther and Ulrike Stein*, 2000
34. **Rotaviruses: *Methods and Protocols*,** edited by *James Gray and Ulrich Desselberger*, 2000
33. **Cytomegalovirus Protocols,** edited by *John Sinclair*, 2000
32. **Alzheimer's Disease: *Methods and Protocols*,** edited by *Nigel M. Hooper*, 2000
31. **Hemostasis and Thrombosis Protocols: *Methods in Molecular Medicine*,** edited by *David J. Perry and K. John Pasi*, 1999
30. **Vascular Disease: *Molecular Biology and Gene Therapy Protocols*,** edited by *Andrew H. Baker*, 1999
29. **DNA Vaccines: *Methods and Protocols*,** edited by *Douglas B. Lowrie and Robert Whalen*, 1999
28. **Cytotoxic Drug Resistance Mechanisms,** edited by *Robert Brown and Uta Böger-Brown*, 1999
27. **Clinical Applications of Capillary Electrophoresis,** edited by *Stephen M. Palfrey*, 1999
26. **Quantitative PCR Protocols,** edited by *Bernd Kochanowski and Udo Reischl*, 1999
25. **Drug Targeting,** edited by *G. E. Francis and Cristina Delgado*, 2000
24. **Antiviral Methods and Protocols,** edited by *Derek Kinchington and Raymond F. Schinazi*, 2000
23. **Peptidomimetics Protocols,** edited by *Wieslaw M. Kazmierski*, 1999
22. **Neurodegeneration Methods and Protocols,** edited by *Jean Harry and Hugh A. Tilson*, 1999

METHODS IN MOLECULAR MEDICINE™

Septic Shock
Methods and Protocols

Edited by

Thomas J. Evans

Imperial College School of Medicine
London, UK

Humana Press **Totowa, New Jersey**

Softcover reprint of the hardcover 1st edition 2000
999 Riverview Drive, Suite 208
Totowa, New Jersey 07512

This publication is printed on acid-free paper. ∞

ANSI Z39.48-1984 (American Standards Institute) Permanence of Paper for Printed Library Materials.

Cover design by Patricia F. Cleary.

Cover illustration: Human renal proximal tubular cells following pro-inflammatory ctyokine injury. The cells have been stained for the epithelial marker cytokeratin (green) with a DAPI nuclear (blue) counter-stain. They have a highly motile phenotype and show extensive filipodia formation. (P. A. Glynne, unpublished observation. *See* Chapter 18.)

For additional copies, pricing for bulk purchases, and/or information about other Humana titles, contact Humana at the above address or at any of the following numbers: Tel: 973-256-1699; Fax: 973-256-8341; E-mail: humana@humanapr.com, or visit our Website at www.humanapress.com

10 9 8 7 6 5 4 3 2 1

Library of Congress Cataloging-in-Publication Data

Septic shock methods and protocols / edited by Thomas J. Evans.
p. cm. -- (Methods in molecular medicine ; 36)
Includes bibliographical references and index.
ISBN 978-1-61737-154-7 e-ISBN 978-1-59259-216-6
1. Septic shock--Research--Methodology. 2. Endotoxins--Research--Methodology. 3. Cytokines--Research--Methodology. I. Evans, Thomas J., 1959. II. Series.
[DNLM: 1. Shock, Septic--immunology. 2. Cell Culture--methods.
3. Cytokines--immunology. 4. Endotoxins--isolation & purification.
5. Nitric Oxide--antagonists & inhibitors. QZ 140 S4796 2000]
RC182.S4S466 2000
6169.047--dc21
DNLM/DLC
for Library of Congress 99-29663
CIP

Preface

Septic shock remains a serious medical condition with high mortality. Despite many advances in intensive care medicine and antibiotic development, this has not changed appreciably in the last 20 years. Frustratingly, over the same period of time, enormous advances have been made in understanding the underlying pathogenic mechanisms of this condition. This has resulted in the development of several novel therapies for septic shock, which, despite excellent theoretical grounds for their efficacy, have failed in altering mortality attributable to sepsis.

The reasons for these failures are multiple, but it is clear that further research is required aimed at increasing our understanding of the basic pathophysiological processes that occur following infection. Research into septic shock draws upon a number of different disciplines, ranging from molecular and cellular biology to physiological measurements on whole animals. *Septic Shock Methods and Protocols* is an attempt to draw together into one volume a number of protocols that are of use in the investigation of the mechanisms of septic shock. I have divided the book into five sections. The first deals with endotoxin, the lipopolysaccharide component of the Gram-negative cell membrane that can mimic many of the features of septic shock. Gram-positive organisms are found increasingly as causes of septic shock, and several aspects of toxins produced from these bacteria are considered in the second section. Cytokines have been a central focus of interest in sepsis research for many years and several aspects of cytokine biology are highlighted in the third section. In the fourth section, methods for studying nitric oxide and other reactive nitrogen intermediates are considered. Finally, the last section describes a variety of methods for studying primary cell cultures, an essential component of developing in vitro methods to study septic shock.

Obviously *Septic Shock Methods and Protocols* cannot provide an exhaustive account of every protocol that might be used in sepsis research. I have therefore carefully chosen those highlights that I believe either are of great intrinsic importance or are poorly covered elsewhere in published protocol manuals. I am very grateful to all the authors who have contributed to this volume. I believe their collected experience is invaluable and I hope that this book will allow both newcomers and those with more experience to apply successfully the techniques needed for their research.

T. J. Evans

Contents

CONTRIBUTORS

JOSEPH S. BECKMAN • *Departments of Anesthesiology, Biochemistry and Molecular Genetics, and Neurobiology and The UAB Center for Free Radical Biology, The University of Alabama at Birmingham, Birmingham, AL*

MAARTEN G. BOUMA • *Department of Surgery, Maastricht University, Maastricht, The Netherlands*

LEE D. K. BUTTERY • *Department of Histochemistry, Imperial College School of Medicine, London, UK*

LORE BRADE • *Research Center Borstel, Center for Medicine and Biosciences, Borstel, Germany*

WIM A. BUURMAN • *Department of Surgery, Maastricht University, Maastricht, The Netherlands*

RAYMOND FOUST III • *Stokes Research Institute and Department of Biochemistry and Biophysics, Children's Hospital of Philadelphia and The University of Pennsylvania, Philadelphia, PA*

THOMAS J. EVANS • *Department of Infectious Diseases, Imperial College School of Medicine, Hammersmith Hospital, London, UK*

JON S. FRIEDLAND • *Department of Infectious Diseases, Imperial College School of Medicine, Hammersmith Hospital, London, UK*

PAUL A. GLYNNE • *Department of Infectious Diseases, Imperial College School of Medicine, Hammersmith Hospital, UK*

CARYN HERTKORN • *Stokes Research Institute and Department of Biochemistry and Biophysics, Children's Hospital of Philadelphia and The University of Pennsylvania, Philadelphia, PA*

DIDIER HEUMANN • *Division of Infectious Diseases, CHUV-1011, Lausanne, Switzerland*

HARRY ISCHIROPOULOS • *Stokes Research Institute and Department of Biochemistry and Biophysics, Children's Hospital of Philadelphia and The University of Pennsylvania, Philadelphia, PA*

PAUL A. KETCHUM • *Associates of Cape Cod, Falmouth, MA*

KEVIN A. KROWN • *Rees-Stealy Research Foundation, San Diego, CA*

MICHAEL LUCHI • *Department of Medicine, Division of Infectious Disease, Kansas University Medical Center, Kansas City, KS*

STUART MALCOLM • *Stokes Research Institute and Department of Biochemistry and Biophysics, Children's Hospital of Philadelphia and The University of Pennsylvania, Philadelphia, PA*

TADEUSZ MALINSKI • *Center for Biomedical Research, Oakland University, Rochester, MI*
DAVID C. MORRISON • *Saint Luke's Hospital/Shawnee Mission Health System; Division of Basic Medical Science, School of Medicine, University of Missouri-Kansas City, Kansas City, MO*
DAVID E. NEWCOMB • *Department of Pathology, University of Michigan, Ann Arbor, MI*
JOANNA PICOT • *Department of Infectious Diseases, Imperial College School of Medicine, Hammersmith Hospital, London, UK*
THOMAS J. NOVITSKY • *Associates of Cape Cod, Falmouth, MA*
JULIA M. POLAK • *Department of Histochemistry, Imperial College School of Medicine, London, UK*
DANIEL G. REMICK • *Department of Pathology, University of Michigan, Ann Arbor, MI*
MANUELA ROGGIANI • *Department of Medicine, Division of Infectious Disease, Kansas University Medical Center, Kansas City, KS*
ROGER A. SABBADINI • *Department of Biology, San Diego State University, San Diego, CA*
PATRICK M. SCHLIEVERT • *Department of Microbiology, University of Minnesota, Minneapolis, MN*
ALEXANDER SHNYRA • *Department of Microbiology, Molecular Genetics and Immunology, Kansas University Medical Center, Kansas City, KS*
DILANI K. SIRIWARDENA • *Moorfields Eye Hospital, London, UK*
SHIRANEE SRISKANDAN • *Department of Infectious Diseases, Imperial College School of Medicine at Hammersmith Hospital, London, UK*
HAJIME TAGORI • *Department of Neonatology, Nagoya City University Medical School, Nagoya, Japan*
CHRISTOPH THIEMERMANN • *The William Harvey Research Institute, St. Bartholomew's and the Royal London School of Medicine and Dentistry, London, UK*
LILIANA VIERA • *Department of Anesthesiology and The UAB Center for Free Radical Biology, The University of Alabama at Birmingham, Birmingham, AL; Departamento de Histologia y Embriologia, Facultad de Medicina, Universidad de la Republica, Montevideo, Uruguay*
JERROLD WEISS • *The Inflammation Program, Departments of Internal Medicine and Microbiology, University of Iowa College of Medicine, Iowa City, IA*
YAO ZU YE • *Department of Anesthesiology and The UAB Center for Free Radical Biology, The University of Alabama at Birmingham, Birmingham, AL*

1

Endotoxin

1

Assay of Endotoxin by *Limulus* Amebocyte Lysate

Paul A. Ketchum and Thomas J. Novitsky

1. Introduction

Horseshoe crabs fight off infectious agents with a complex array of proteins present in amebocytes, the major cell type in their hemolymph. These amebocytes contain both large and small granules *(1)*. When exposed to bacteria or other infectious agents the amebocytes release proteins into their surroundings by exocytosis. The small granules of *Limulus* amebocytes contain antibacterial proteins, including polyphemusins and the big defensins *(2)*. The large granules contain the *Limulus* anti-lipopolysaccharide factor (LALF) and the clot-forming group of serine protease zymogens. Exocytosis is initiated by the reaction of amebocytes with lipopolysaccharide (LPS) from Gram-negative bacteria or other microbial components. LPS is also called endotoxin because it is found in the outer membrane of the gram-negative bacterial cell wall. A solid clot forms in response to the lipid A portion of LPS, thereby walling off the infection site or preventing the loss of blood when the animal is damaged physically *(3)*.

The clot-forming cascade of serine proteases is the basis for the *Limulus* amebocyte lysate (LAL) assay for endotoxin (**Fig. 1**). Factor C is activated autocatalytically by LPS, which in turn activates factor B, which then activates the proclotting enzyme *(4)*. The activated clotting enzyme cleaves coagulogen to coagulin, which forms the firm clot. Clot formation was the basis for the first LAL assay for endotoxin *(5)*. The LAL assay has replaced other tests (e.g., the rabbit pyrogen test) in part because the LAL cascade amplifies the initial signal (LPS) greatly, permitting the detection of picogram quantities of LPS. Clot formation can also be initiated by (1→3)-β-D-glucan (**Fig. 1**) from fungal cell wall, (*see* **Note 7; refs.** ***6,7***).

From: *Methods in Molecular Medicine, Vol. 36: Septic Shock*
Edited by: T. J. Evans © Humana Press Inc., Totowa, NJ

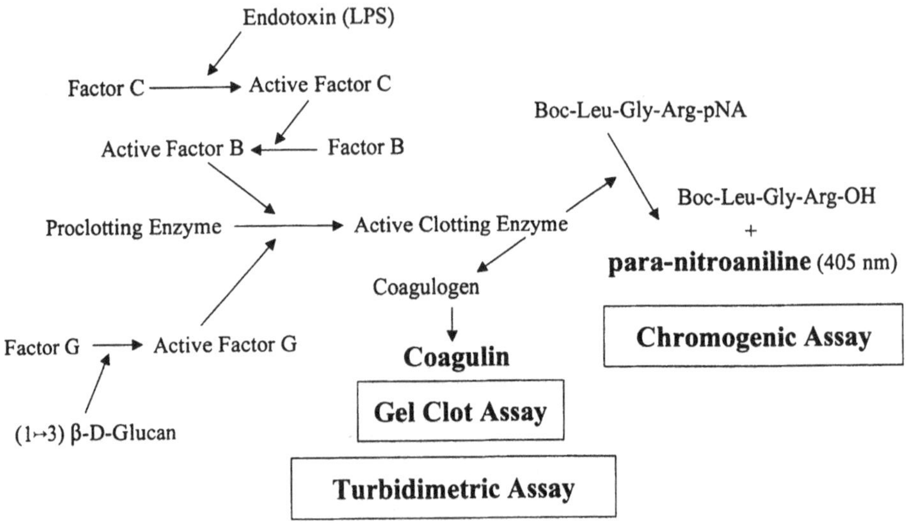

Fig. 1. The *Limulus* blood clotting cascade.

1.1. LAL Method for Measuring Endotoxin

Since the LAL gel-clot assay was first approved in 1977 by the Food and Drug Administration (FDA) for detecting endotoxin, manufacturers have developed two additional LAL methods. The turbidimetric LAL method is an adaptation of the gel-clot to instrumental analysis ***(8)***. Turbidity is monitored as an increase in light scattering caused by clot initiation. The LAL reagent is specially formulated to be incubated in a test-tube reader such as the LAL-5000 ***(9)***, which measures the turbidity of each tube with time. The computer-based software determines the time for the reaction to reach a predetermined onset optical density (OD). The log of the onset time is linearly related to the log of the endotoxin concentration ***(9)***.

The chromogenic LAL method utilizes a peptide substrate that turns yellow when hydrolyzed by the proclotting enzyme ***(10)***. One example is the peptide substrate Boc-Leu-Gly-Arg-*p*-nitroanilide shown in **Fig. 1**. Activated proclotting enzyme cleaves the chromophore from the arginine, releasing the yellow-colored *p*-nitroaniline (*p*NA). In the normal end-point assay, the amount of pNA released is determined after a prescribed 37°C incubation by reading the OD or absorbance at 405 nm. This reagent can also be used in a kinetic assay where the time required to attain an onset at OD_{405} (usually 0.03–0.1) is related to the endotoxin concentration.

The LAL assay for blood endotoxin is composed of three basic parts: sample collection and handling; extraction of the blood/serum sample; and testing the extracted sample with the chromogenic LAL assay. Both the chromogenic end-point assay and the turbidimetric assays are used to detect endotoxin in body fluids; however, here we describe the chromogenic method. For a recent review of the literature see Novitsky *(11)*.

1.2. Interfering Substances in Blood

Animal blood contains soluble enzymes, antibodies, LPS binding proteins, and HDL that interfere with the detection of endotoxin by LAL assays. The serine proteases present in blood can act on the chromogenic substrate in the absence of the LAL reagent and must be inactivated. Moreover, humans possess sophisticated mechanisms for binding, transporting, and eventually processing LPS to remove it from the circulation. LPS binding protein, cationic antibacterial proteins, and bacterial permeability-increasing protein are examples of serum proteins that bind LPS and interfere with endotoxin measurement. The degree of interference varies among patient sera as demonstrated by Warren et al. *(12)* through studies on the plasma samples from blood donors. Some individuals also have high concentrations of serum antibodies against endotoxin *(13)* capable of neutralizing its biological effects. Two methods are available to deal with serum-protein interference: the heat dilution method *(14,15)*, and the acid treatment described in **Subheading 3.2.2.** *(16)*.

Certain blood samples have a yellow color whose absorbance interferes with measuring *p*NA at 405 nm. This interference is avoided by diazo-coupling the *p*-NA, thus forming a purple complex that absorbs at 540–550 nm with a threefold higher extinction coefficient than *p*NA. The diazo-coupling method is useful in the chromogenic endpoint LAL assay *(17)*.

2. Materials

2.1. Equipment Required

1. The end-point chromogenic LAL method requires a microplate reader with a 545-nm filter for measuring diazo-coupled *p*NA and a 640-nm filter for eliminating background interference. The plate reader is connected to a computer with a software package suitable for analyzing the results of the LAL assay.
2. Incubating the plate at 37°C requires either a temperature-controlled microplate reader or a microplate block incubator. Either a water bath or a heating block at 37°C is used during the blood-extraction procedure.
3. A clinical centrifuge capable of 1300–1500*g* is used to prepare blood plasma and perform the blood-extraction protocol.

2.2. Laboratory Reagents and Materials

1. All materials used directly in the assay must be essentially free of endotoxin.
   a. LAL reagent-grade water (LRW), glass pipets (Fisher, Pittsburgh, PA), certified microtiter plates, Rainin pipet tips, and Eppendorf combitips are recommended.

b. Depyrogenated blood extraction tubes (10 × 75-mm) and any other glassware is wrapped in aluminum foil and baked at 240°C for at least 4 h.
c. Purple-top ethylenediaminetetraacetic acid (EDTA) Vacutainer (Fisher) tubes are used for blood collection. Heparin Vacutainer tubes certified to be endotoxin-free may be substituted.
d. The gloves worn during blood handling and performing the assay must be powder-free, because the powder contains endotoxin and can contaminate the assay.

2. The chromogenic LAL reagent kit with endotoxin standard and diazotization reagents is available from Associates of Cape Cod (Falmouth, MA). This kit contains Pyrochrome LAL, Pyrochrome Buffer, endotoxin standard and the diazo-coupling reagents. Other manufactures supply the chromogenic LAL reagent suitable for the assay and an endotoxin standard. If not purchased, the diazotization reagents are made according to information in **Table 1**.
3. The blood-extraction reagents are 0.5% Triton X-100 prepared in LRW, 1.32 *N* HNO_3 diluted from concentrated HNO_3 in LRW, and 0.55 *N* NaOH prepared by dissolving solid NaOH in LRW. These reagents are stable at room temperature.

2.3. LAL Product Insert

1. The product insert provided with each lot of LAL reagent contains valuable information on how to reconstitute the LAL reagent, storage of the reconstituted LAL, testing methods, volumes of reagent to use, sensitivity of the reagent, and recommended endotoxin standards. Because LAL is a biological product, the conditions of storage and stability of the reconstituted reagent are critical to success.
2. LAL reagents are licensed by the FDA and other regulatory bodies for detection of endotoxin in pharmaceutical preparations. They are not licensed for the detection of endotoxin in blood and other body fluids. When used for this purpose, the results are for research use only.

2.4. Endotoxin Standard

1. The reference standard endotoxin (RSE) is made from *Escherichia coli* 0113 and known as EC-6. Other endotoxin standards are related to RSE and their potency documented in a certificate of analysis (Control Standard Endotoxin; CSE).
2. The quantity of endotoxin is recorded as an endotoxin unit (EU): one EU is equivalent to 100 pg of RSE. Endotoxin is routinely reported as EU/mL.
3. The endpoint assay with diazo-coupling is sensitive over the range of 0.25–0.015 EU/ml. Reconstituted endotoxin standards are stable for >1 wk at 4–8°C.
4. Because endotoxin forms micelles and binds to glass surfaces, solutions of reconstituted endotoxin are vortexed for 5 min or longer. Each dilution made in a test tube should be vortexed for 0.5–1.0 min before use or further dilution. Endotoxin standards are usually diluted in LRW or in diluted (1/10) pyrogen-free

Table 1
Reagents for the Chromogenic LAL Assay

	Composition	Storage
Blood Extraction Reagents		
Nitric acid	1.32 *N*	Room temperature
Triton X-100	0.5% (v/v)	Room temperature
NaOH	0.55 *N*	Room temperature
Pyrochrome Reagents		
Pyrochrome (lyophilized)	Reconstitute with 3.2 mL buffer	4–8°C
Pyrochrome reconstitution buffer	0.2 *M* Tris HCl pH 8.0 (23°C)	Room temperature
Endotoxin		
Endotoxin standard (lyophilized)	Make 0.25 EU/mL	4–8°C
LAL reagent water		Room temperature
Diazo-coupling reagents		
Sodium nitrite (lyophilized)	0.417 mg/mL in 0.48 *N* HCl (below)	Room temperature
Hydrochloric acid	0.48 *N*	Room temperature
Ammonium sulfamate (lyophilized)	3 mg/mL	Room temperature
n[1-naphthyl]-ethylenediamine dihydrochloride (NEDA)	0.7 mg/mL LRW	Room temperature

plasma. Dilutions can be performed in pyrogen-free test tubes or in the microtiter plate.

3. Methods

3.1. Sample Collection and Handling

1. Blood samples can be drawn from lines or fresh sticks into the Vacutainer purple-top tube *(18)*. The sample is placed in ice and transported to the laboratory for making plasma (*see* **Note 1**).
2. When present, blood endotoxin levels in sepsis patients tend to remain elevated over a period of days *(18)*. Unless one is looking for a specific event, timing of blood collection within the 12 h following onset of sepsis is not a critical factor in detecting blood endotoxin.
3. Plasma samples can be subdivided and tested before freezing, or stored at –80°C for months before doing the assay (*see* **Note 2**). Storage at –80°C and transportation on dry ice is advised.

3.2. Protocol for the Chromogenic LAL Method for Endotoxin Detection

3.2.1. Setting Up the LAL Assay

Set up the LAL assay in a biosafety cabinet or laminar flow hood. If this is not possible, the technician should take special precautions to ensure that the work space is free of dust and the reagents and materials do not become contaminated. Perform the assay in an isolated area with restricted traffic and minimal interference. *Do not* lean over the microplate when adding samples to the wells. Keep the microplate lid closed unless adding samples or performing dilutions. Always use aseptic techniques when pipeting.

3.2.2. Preparing the Blood Sample

Wear nonpowdered gloves and observe the safety regulations for blood handling as directed by your institution. These instructions apply for each blood sample.

1. Place two sterile endotoxin-free 10 × 75-mm test tubes on ice and label them "A" for acidification and "B" for neutralization.
2. Add 200 µL of nitric acid and 200 µL of Triton X-100 to the "A" tube (use within 30 min, do not store mixture).
3. With a separate pipet tip, add 200 µL of sodium hydroxide to the "B" tube.
4. Thaw frozen samples at room temperature, then vortex them for 1 min prior to transferring 100 µL of blood to tube "A." Cover the tube with the nonexposed side of parafilm and vortex for 30 s.
5. Immediately incubate tube "A" at 37°C for 5 min.
6. Again vortex tube "A" for 30 s and centrifuge at 1300–1500*g* for 5 min. Remove tube "A" and place on ice.
7. Using an endotoxin-free pipet tip, transfer 200 µL of the supernatant fluid from tube "A" to tube "B" (containing sodium hydroxide).
8. Vortex tube "B" for 5 s then store on ice until assayed. This represents a 1/10 dilution of the blood sample.

3.2.3. Setting up the Microplate

Before thawing the samples, turn on the plate-heating block or the plate-heating reader, and prepare the tube heater.

1. Set up the OD plate reader as follows:

Temperature	37°C
Automix	Off
Wavelength	540 or 550 nm
Background wavelength	630 nm
Calibration	on

2. Using the appropriate plate-reader software (e.g., Molecular Devices Softmax), designate the microplate wells that will be used for OD readings. The assay requires duplicate wells for the water blanks (negative controls) and for each endotoxin concentration in the range of 0.25–0.015625 EU/mL (positive control and standard). By not using the outside wells on the plate, one reduces the probability of chance contamination when handling the plate.
3. Test sample dilutions ranging from 1 : 10–1 : 320. To conserve reagents, one can do dilutions of 1:20, 1:40, and 1:80 and obtain results for all but the highest endotoxin concentrations.

3.2.4. Control Standard Endotoxin

Prepare the control standard endotoxin at 0.25 EU/mL. Reconstitute the dried CSE with LRW, then vortex for at least 1 min. Endotoxin can be stored at room temperature until used the same day.

3.2.5. Chromogenic LAL Reagent

Prepare the Chromogenic LAL reagent as recommended by the manufacturer (*see* product insert and **Note 3**). Be sure to use aseptic technique and endotoxin-free buffer or the water supplied with the reagent. Swirl gently, then cover with the unexposed surface of parafilm and place on ice. Most formulations should be used within 2 h of reconstitution. Some can be frozen (stored for >1 wk), thawed, and used without problems (*see* manufacturers' product insert).

3.2.6. Additions of Samples and Standards to Microplate

Add 50 µl of LRW to the blank wells and to each well a 1:1 dilution is to be made. Now add 50 µL of the highest concentration of endotoxin (0.25 EU/mL) to the empty well and to the next well containing 50 µL of LRW, thus making a 1 : 1 dilution (0.125 EU/mL). Continue the dilution series first for the standard by mixing with the pipet then transferring 50 µL to the next well. After mixing the last dilution, discard 50 µL to waste. Now each well has 50 µL of sample or water (blank). Repeat the process for each sample. All dilutions are assayed in duplicate.

3.2.7. Addition of Chromogenic LAL Reagent

Add a pipet tip to the Eppendorf combitip and rinse once with LRW by filling the pipet and expelling the water to waste. Fill the washed pipet with chromogenic LAL and set to dispense 50 µL into each well. Do not touch the samples in the plate with the pipet tip. Add from the lowest concentration (highest dilution) to the highest concentration. For best results this step should be done quickly without splashing. Replace the cover, then mix either by shaking gently on a flat surface or use a mechanical mixing platform (may be in plate reader) for 10 s and incubate at 37°C for a prescribed time (usually 25–30 min).

3.2.8. Stopping the Reaction

While the assay is incubating, prepare the diazo-reagents (**Table 1**). At the end of the incubation period, place the plate on the counter and add 50 μL of the sodium nitrite dissolved in the 0.48 *N* HCl to stop the reaction (*see* **Note 4**). Make this addition quickly in the same sequence as the addition of the chromogenic LAL (*see* **Subheading 3.2.7.**). Next add 50 μL of the ammonium sulfamate to each well and gently mix. Finally add 50 μL of the n[1-naphthyl]-ethylenediamine dihydrochloride (NEDA) solution and allow the color to develop at room temperature for 5 min.

3.2.9. Reading the Assay and Determining Endotoxin Concentration

Insert the microplate in the plate reader with preset parameters as described in **Subheading 3.2.3.** The instrument will read the ODs of the wells containing the blanks, standards, and unknowns. The OD_{545} of the standards are plotted against the endotoxin concentration (EU/mL). The blanks should be <0.100 and the absolute *r* for the line should be ≥ 0.980. Many software programs will plot the standard curve (with or without blanks subtracted) and calculate the endotoxin concentration of the unknowns relative to that internal standard.

4. Notes

1. Although whole blood can be used with the acid-extraction method, there appears to be no advantage to using whole blood. Removal of cell mass from whole-blood concentrates the endotoxin in the smaller plasma volume, so plasma samples will contain more endotoxin/volume than whole blood ***(18)***.
2. In one study of 354 samples, the assay was repeatable on frozen samples performed at different sites; however, certain fresh samples (7%) tested before freezing registered more endotoxin than the frozen/thawed samples tested at a later date. Therefore freezing may affect the recovery of endotoxin in certain samples ***(18)***.
3. Variations on this chromogenic LAL protocol are used by certain manufacturers. For example, the COATEST Plasma reagent (Chromogenix, Milan, Italy) is designed for a two-step protocol in which the chromogenic substrate is separate from the lysate. In the two-step procedure, the reconstituted LAL reagent is added to the preheated samples and then incubated for a short time (5–14 min depending on the endotoxin range being tested). Next, the buffered chromogenic substrate is added to each well and the plate again incubated at 37°C for 5 or 8 min (product insert) before stopping the reaction.

4. The reaction can be stopped by adding acetic acid (20%) to each well. With this method, the end product is *p*NA, whose concentration is determined at 405 nm.
5. The kinetic method broadens the sensitivity range to 1.0–0.05 EU/mL using the chromogenic reagent and to 10–0.005 EU/mL using the turbidimetric assay (before factoring in the sample dilution).
6. False positive results with the chromogenic LAL: Blood-borne interfering substances that cause a false positive with this assay are known *(18)*. The plasma of patients treated with certain sulfa antimicrobial agents can give a false-positive reaction when the diazo-coupling reagents are used. Sulfamethoxazole, sulfisoxazole, sulfapyridine, and sulfanilamide form diazo complexes that absorb at 545 nm. Samples from patients treated with sulfa drugs should be tested with the diazo-coupling reagents as a control before testing with the chromogenic LAL reagent.
7. Many fungi contain β-D-glucans as components of their cell walls *(19)*. These carbohydrates activate factor G of the LAL cascade (**Fig. 1**), resulting in activation of the proclotting enzyme. This in turn cleaves *p*NA from the peptide substrate giving a positive reaction. Endotoxin-specific reagents are available from Seikagaku Corp. (Tokyo, Japan) for determining this type of interference.

References

1. Toh, Y., Mizutani, A., Tokunaga, F., Muta, T., and Iwanaga, S. (1991) Morphology of the granular hemocytes of the Japanese horseshoe crab *Tachypleus tridentatus* and immunocytochemical localization of clotting factors and antimicrobial substances. *Cell Tissue Res.* **266,** 137–147.
2. Iwanaga, S., Kawabata, S., and Muta, T. (1998) New types of clotting factors and defense molecules found in horseshoe crab hemolymph: their structures and functions. *J. Biochem.* **123,** 1–15.
3. Levin, J. and Bang, F. B. (1964) The role of endotoxin in the extracellular coagulation of *Limulus* blood. *Bull. Johns Hopkins Hosp.* **115,** 265–274.
4. Iwanaga, S., Miyata, T., Tokunaga, F., and Muta, T. (1992) Molecular mechanism of hemolymph clotting system in *Limulus. Thrombos. Res.* **68,** 1–32.
5. Levin, J., Tomasulo, P. A., and Oser, R. S. (1970) Detection of endotoxin in human blood and demonstration of an inhibitor. *J. Lab. Clin. Med.* **75,** 903–911.
6. Morita, T., Tanaka, S., Nakamura, T., and Iwanaga, S. (1981) A new (1→3) -β-D-glucan-mediated coagulation pathway found in *Limulus* amebocytes. *FEBS Lett.* **129,** 318–321.
7. Roslansky, P. F. and Novitsky, T. J. (1991) Sensitivity of *Limulus* amebocyte lysate (LAL) to LAL-reactive glucans. *J. Clin. Microbiol.* **29,** 2477–2483.
8. Fink, P. C., Lehr, L., Urbaschek, R. M., and Kozak, J. (1981) *Limulus* amebocyte lysate test for endotoxemia: investigations with a femtogram sensitive spectrophometric assay. *Klin. Wochenschr.* **59,** 213–218.
9. Remillard, J., Gould, M. C., Roslansky, P. F., and Novitsky, T. J. (1987) Quantitation of endotoxin in products using the LAL kinetic turbidimetric assay, in *Detection of Bacterial Endotoxins with the* Limulus *amebocyte lysate test*

(Watson, S., Levin, J., and Novitsky, T. J., eds.), Alan R. Liss, New York, pp. 197–210.
10. Iwanaga, S., Morita, T., Harada, T., Nakamura, S., Niwa, M., Takada, K., Kimura, T., and Skakibara, S. (1978) Chromogenic substrates for horseshoe crab clotting enzyme: its application for the assay of bacterial endotoxins. *Haemostasis* **7,** 183–188.
11. Novitsky, T. J. (1994) *Limulus* amebocyte lysate (LAL) detection of endotoxin in human blood. *J. Endotoxin Res.* **1,** 253–263.
12. Warren, H. S., Novitsky, T. J., Ketchum, P. A., Roslansky, P. F., Kania, S., and Siber, G. R. (1985) Neutralization of bacterial lipopolysaccharide by human plasma. *J. Clin. Microbiol.* **22,** 590–595.
13. Greisman, S. E., Young, E. J., and Dubuy, B. (1978) Mechanism of endotoxin tolerance. VII. Specificity of serum transfer. *J. Immunol.* **111,** 1349–1360.
14. Cooperstock, M. S., Tucker, R. P., and Baublis, J. V. (1975) Possible pathogenic role of endotoxin in Reye's syndrome. *Lancet* **1,** 1272–1274.
15. Roth, R. I., Levin, F. C., and Levin, J. (1990) Optimization of detection of bacterial endotoxin in plasma with the *Limulus* test. *J. Lab. Clin. Med.* **116,** 153–161.
16. Tamura, H., Tanaka, S., Obayashi, T., Yoshida, M., and Kawai, T. (1991) A new sensitive method for determining endotoxin in whole blood. *Clin. Chim. Acta* **200,** 35–42.
17. Tamura, H., Tanaka, S., Obayashi, T., Yoshida, M., and Kawai, T. (1992) A new sensitive microplate assay of plasma endotoxin. *J. Clin. Lab Anal.* **6,** 232–238.
18. Ketchum, P. A., Parsonnet, J., Stotts, L. S., Novitsky, T. J., Schlain, B., Bates, D. W., and Investigators of the AMCC SEPSIS Project. (1997) Utilization of a chromogenic *Limulus* amebocyte lysate blood assay in a multi-center study of sepsis. *J. Endotoxin Res.* **4,** 9–16.
19. Obayashi, T., Yoshida, M., Mori, T., Goto, H., Yasuoka, A., Iwasaki, H., Teshima, H., Kohno, S., Horiuchi, A., Ito, A., Yamaguchi, H., Shimada, K., and Kawai, T. (1995) Plasma (1→3)-β-D-glucan measurement in diagnosis of invasive deep mycosis and fungal febrile episodes. *Lancet* **345,** 17–20.

2

Preparation of Endotoxin from Pathogenic Gram-Negative Bacteria

Alexander Shnyra, Michael Luchi, and David C. Morrison

1. Introduction

Endotoxins have been recognized for decades as important structural components of the outer cell wall/cell membrane complex of Gram-negative microorganisms. These chemically heterogeneous macromolecular structures were recognized very early on to consist of lipid, polysaccharide, and protein, and to have the capacity to induce deleterious pathophysiological changes when administered either systemically or locally to a wide variety of experimental laboratory animals. The recognition of the very significant disease-causing potential of these interesting microbial constituents provided a sound conceptual basis for studies directed at the isolation, purification, and detailed chemical characterization of the active constituent(s). It is perhaps not particularly surprising, therefore, that there are now numerous methods and modifications of methods, that have been published in the scientific literature describing various approaches that have been employed for the extraction and purification of endotoxin from bacteria. It would be beyond the scope of this chapter to describe in detail all of these various methods. Therefore, we shall provide only a brief historical perspective of the evolution of different methodologies. We will then focus upon a more detailed discussion of those that will ultimately serve the investigative purposes of most researchers interested in isolating and purifying endotoxins.

It would be of value at the outset to begin with a definition of exactly what is meant in this chapter when referring to the terms "endotoxin" and "lipopolysaccharide" (LPS). Although these two terms are often used interchangeably, it is important to note that they are both functionally and biochemically distinct

From: *Methods in Molecular Medicine, Vol. 36: Septic Shock*
Edited by: T. J. Evans © Humana Press Inc., Totowa, NJ

entities. Almost by definition, endotoxin can refer to any microbial extract that is enriched for an activity that will induce, either in vitro or in vivo, some or all of the pathophysiological characteristic manifestations of Gram-negative microbes. And, as will be pointed out below, there is no *a priori* requirement that endotoxin be a highly purified substance. In rather marked contrast to this, LPS is a chemically defined entity, usually consisting of a characteristic lipid (lipid A) covalently linked to varying amounts of polysaccharide, and free of other contaminating microbial constituents. The very fundamental feature of varying chemical structures embodied in the latter requires that the term LPS be used to describe a class of biochemically active microbial constituents rather than a single well-defined structure.

Among the first major investigators to address the question of the identity of endotoxin was Andre Boivin, who employed a cold trichloroacetic acid (TCA) procedure to Gram-negative microbes *(1)*. The resulting relatively impure extract nevertheless retained many of the early classical endotoxic biological properties recognized to be characteristic of endotoxin. Such preparations, in addition to containing lipid or polysaccharide were also known to contain substantial amounts of microbial proteins, although the extent to which these proteins were physically associated with the lipids/carbohydrates was not determined. Endotoxins extracted by such procedures are still available from at least one commercial distributor, although both significant refinements in purification of the endotoxically active components and an increased appreciation of the potential role of other microbial factors to expression of biological activity have impacted upon their general use by investigators. Nevertheless, in some circumstances relatively impure preparations of endotoxin containing other microbial constituents may actually be perceived as an advantage for a given investigative purpose and, under those circumstances, Boivin-type TCA-extracted materials might be the endotoxin of choice.

The seminal studies by Westphal, Luderitz, and their collaborators established what is now considered by most endotoxin researchers to be the gold standard for the isolation and purification of relatively chemically homogeneous preparations of endotoxin *(2)*. Perhaps equally noteworthy, however, is the fact that development of this methodology ultimately led to the discovery of the fundamental essence of the endotoxic principal of the Gram-negative microbe. These investigators used a hot aqueous phenol procedure to isolate essentially protein-free preparations of LPS, covalent conjugates of lipid and polysaccharide. The very clear demonstration by Westphal and Luderitz that mild acid hydrolysis of LPS resulted in the selective cleavage of a very acid labile bond in LPS would then allow the generation of an aqueous insoluble white precipitate that embodied virtually all of the biologically active endotoxic activities of the original LPS-enriched phenolic extracts, first established the overriding importance, not only of LPS, but also its covalently linked lipid part in endotoxin-initiated host responses *(3)*. Westphal and Luderitz termed this lipid fraction as lipid A (denoted to mean the covalently associated lipid) that served to distinguish it from the other lipid fraction found in the hot phenolic extracts (lipid B) that could be extractable into nonaqueous polar solvents without acid hydrolysis pretreatment. The hot phenol-water extraction procedure has been employed by numerous investigators for purification of endotoxic LPS from a variety of microorganisms. Because of its broad and almost universal usage, we shall describe this basic method as well as the adoption of common refinements on it for the preparation and purification of LPS.

The LPS of many, but not all, Gram-negative bacteria is now well recognized to consist of three major domains: lipid A; the core or oligosaccharide domain; and the polysaccharide chain of repeating units of O antigen *(4)*. LPS, therefore, is representative of a class of amphiphilic macromolecules with up to seven hydrophobic fatty acyl groups in the lipid A domain and with a hydrophilic polysaccharide constituent possessing negatively charged phosphate and carboxy groups. Although most *Enterobacteriaceae* microorganisms manifest this complete LPS molecule with a lipid A–core–O polysaccharide structure, some nonenterobacterial species, such as *Neisseria*, *Haemophilus influenzae*, *Bordetella pertussis*, *Acinetobacter* as well as a variety of well-characterized mutant strains of *Enterobacteriaceae* are deficient in synthesis of O-specific chain and parts of the core. As a consequence, such microbes synthesize only lipid A and either part or all of the core region of the LPS macromolecule. These bacteria were recognized relatively early on to form so-called rough colonies on agar plates. Therefore, LPS isolated from such bacteria was originally termed R—or rough—chemotype LPS, to distinguish them from S—or smooth—chemotype LPS manifested by bacteria that grow in smooth colonies because of a complete LPS structure. In the scientific literature, such LPS preparations acquired the name lipooligosaccharides (LOS) although the term R-chemotype LPS is still in common usage, particularly when referring to such LPS preparations from enteric microorganisms *(5)*.

The absence of a chemically defined O-polysaccharide domain in R-LPS/LOS results in a shift towards more hydrophobic physicochemical properties of the LPS macromolecule. Consequently, extraction of R-LPS into an aqueous hydrophilic phase by means of standard phenol-water extraction procedures generally resulted in relatively low yields of R-LPS. To overcome this problem, a hydrophobic extraction procedure based on the use of phenol-chloroform-light petroleum ether (PCP) was developed originally by C. Galanos in the laboratory of O. Westphal and O. Luderitz *(6)*. Because of both relatively mild extraction conditions (e.g., room temperature) and the hydrophobic nature of the extraction mixture, the development of this PCP procedure resulted in yields of R-LPS preparations that contain only traces of contaminating RNA, DNA, and protein. The broad utility of this method for extraction and purification of R-LPS has also placed this method among the most common techniques for purification of LPS in which the standard hot phenol water technique has not proven appropriate, and we shall, therefore, also describe this method in detail in this chapter.

1.1. General Considerations

There are several general considerations that should be addressed prior to undertaking the purification of endotoxic lipopolysaccharides from Gram-negative microbes. These include decisions regarding the microorganism from which the endotoxin will be extracted, the amount of purified material that will be required to totally fulfill the requirements of the investigator, whether such material is available commercially already, and the degree of purity/homogeneity that will be required. The following paragraphs will briefly address each of these specific issues.

Regarding the microorganism itself, it is important to decide initially whether the microbe manifests an R or S phenotype as this will influence whether or not the hot phenol-water procedure or the phenol-chloroform-petroleum ether method is adopted. Usually this information is known. However, if it is unclear or if the possibility exists that the phenotype may vary depending upon growth conditions, it may be necessary to try both approaches to ascertain empirically which method would yield more optimal results. In many instances, additional information can be obtained by carrying out sodium dodecyl sulfate—polyacrilimide gel electrophoresis (SDS-PAGE) analysis of whole microbe extracts using the protease K procedure described by Hitchcock and Brown *(7)* followed by silver-staining of the electrophoresed LPS using the technique of Tsai and Frasch *(8)*. However, although highly effective, these procedures are not routinely recommended for those whose laboratory is not already set up to do these types of studies, and expert advice should be sought before undertaking them.

A second decision regards the growth conditions should be employed to prepare the starting material for the preparation of the LPS. Experimental evidence indicates that the actual chemical composition of the LPS can vary, at least to some extent, depending upon the phase of growth of the microorganism. For example, many bacteria in the logarithmic phase of growth manifest less O-antigen polysaccharide relative to lipid A content than do the same organisms in the late logarithmic or stationary phase of growth (*see* **ref. *9***). Other microorganisms that synthesize primarily R-LPS may express different structures on their abbreviated core oligosaccharide depending on the growth temperature of the cultures (e.g., *Yersinia* LPS at 30° vs 37°C, R.R. Brubaker and D.C. Morrison, unpubl.). Whereas all of the variables that might influence structural determinants have not been investigated in detail, a good rule of thumb would be to use conditions as close as possible to those that might be anticipated in the real world to prepare the bacteria for extraction.

A third factor that merits consideration is the anticipated yields of purified LPS and the relationship of yield to anticipated demands. In general, it can be estimated that the LPS component of many microorganisms constitutes approx 5–10% of the total dry weight of the bacteria. Wet weight of bacteria freshly harvested from in vitro are approx 1 mg of packed cells per 5×10^8 bacteria and total dry weight approx 25% of that value. Thus, a liter of late-logarithmic-phase cells will contain approx 1 g of wet weight packed cells, 250 mg of dried bacterial mass, and approx 12–25 mg of LPS content. Assuming an average yield of 25–50% of the total available LPS, therefore, one might estimate that a reasonable expectation of LPS from a liter of bacterial culture would be somewhere between 5 and 10 mg of purified material. When scaling up to much larger volumes and large-scale purification efforts, yields are invariably somewhat less than linearly proportional. Nevertheless, these general guidelines are not unrealistic as first approximations.

A final major consideration that needs to be addressed is the degree of purity that will be required for the investigator to pursue the proposed studies. Although it is relatively straightforward to prepare LPS from cultures of Gram-negative microbes that are enriched for the endotoxic LPS constituent, it is a much greater challenge to prepare LPS that is absolutely devoid of all other microbial constituents. In this respect, potential contaminants would include (depending upon starting material and method of extraction), capsular polysaccharide, nucleic acid, and protein, particularly outer membrane proteins that are well recognized for their potential high-affinity binding to LPS ***(10)***. Although the presence of these contaminants can, for many purposes, be irrelevant, it is important to keep in mind that many contaminants do manifest their own biological activities that, in the past, have complicated the interpretation of experimental data (*see* **refs. *11,12***).

2. Materials

2.1. Growth of Bacteria

1. Magnetic hot plate stirrer with stirring bar.
2. Tryptone and yeast extract or Luria-Bertani (LB) broth (Difco Laboratories, Detroit, MI).
3. NH_4Cl, Na_2HPO_4, KH_2PO_4, and Na_2SO_4 that meet American Chemical Society (ACS) specifications.

4. Disposable sterile plastic tubes (15 mL) (Becton Dickinson, San Jose, CA).
5. Erlenmeyer flasks (2L) (Kimble Glass, Vineland, NJ).
6. Orbital rotary shaker platform with temperature control.
7. High-volume centrifuge and 250-mL polycarbonate centrifuge tubes with caps.

2.2 Extraction of Bacterial Lipopolysaccharide: Phenol-Water Extraction

1. Crystalline phenol (*see* **Note 1**).
2. RNase.
3. DNase.
4. Proteinase K.
5. Refrigerated centrifuge.
6. Hot plate/stirrer, stir bars.
7. Thermometer.
8. Glassware (50-mL graduated cylinder, 50- and 200-mL beakers).
9. Glass beaker (2000 mL) or glass tray.
10. Glass pipets.
11. Glass tubes for centrifugation.
12. Dialysis tubing, 12,000–14,000 molecular weight cutoff (MWCO).

2.3. Extraction of Bacterial Lipopolysaccharide: Phenol-Chloroform-Petroleum Ether

1. Round-bottom or short conical-bottom glass centrifuge tubes (50 mL) (Kimble Glass).
2. Ethanol, acetone, diethyl ether, chloroform, light petroleum ether (boiling range 40–60°C): all of ACS grade.
3. Ultra-Turrax laboratory homogenizer, IKA Works, (VWR Scientific Products, South Plainfield, NJ).
4. Rotary evaporator, R-114 Series, Brinkmann (VWR Scientific Products).
5. Ultrasonic bath, Fisher Ultrasonic Cleaner (Fisher Scientific, Pittsburgh, PA).
6. Dialyzing tubing, molecular weight (MW) cutoff 12,000–14,000 (Spectra/Por) (Spectrum Medical Industries, Laguna Hills, CA).

3. Methods

3.1. Growth of Bacteria

Unless otherwise stated, or unless very special growth conditions are required, the following very general growth medium and culture conditions can be employed in the preparation of microbes for subsequent extraction and purification of LPS.

1. To prepare the growth medium, dissolve 10 g tryptone, 5 g yeast extract, 2.5 g NH_4Cl, 15 g Na_2HPO_4, 6 g KH_2PO_4, and 0.5 g NA_2SO_4 in 1 L of deionized water

by heating the mixture with constant stirring on a magnetic hotplate stirrer. Alternatively, bacteria can be grown in LB broth (Difco Laboratories) prepared according to the manufacturer protocol.
2. Dispense the medium into tubes and flasks and autoclave for 15 min at 121°C.
3. Transfer a 1-μL disposable sterile loop of stock bacteria (keep frozen at –70°C) into a tube with 10 mL of tryptone-yeast extract medium and incubate overnight at 37°C (*see* **Note 2**).
4. On the following day, inoculate 10 mL of this culture into 1.5 L of the medium in a 2-L Erlenmeyer flask and grow the bacteria on an orbital shaker (150–200 rpm at 37°C) to the late logarithmic phase in submerged cultures for 36 h at 37°C.
5. Harvest microorganisms by dispensing volumes of 200 mL each into the 250-mL centrifuge tubes and centrifuging at 9000*g* for 15 min. Discard the centrifuge supernatants and add additional bacteria plus growth medium until all of the bacteria have been pelleted by centrifugation.
6. Resuspend the bacterial pellets in a small volume (e.g., 10–20 mL) of sterile pyrogen-free water by vigorous pipetting, vortexing and mixing, and combine all of the bacterial pellets into one suspension in one of the centrifuge tubes. Fill to 200 mL with pyrogen-free distilled water and wash by centrifugation using the conditions described above at least one more time. You can estimate the total approximate number of organisms by making a 1 : 1000 dilution of the final dispensed pellet and determining the light-scattering capacity in a standard spectrophotometer at 650 nm using the conversion figure of 0.80 absorbance units/cm = 5.0×10^8 cfu/mL. This preparation, or some multiple or fraction of it, can, in general, serve as the starting material for the extraction and purification of LPS.

3.2. Extraction of Bacterial Lipopolysaccharide: Phenol-Water Extraction

The purification of smooth LPS from whole Gram-negative bacteria by the phenol-water extraction procedures is essentially unchanged from that originally reported by Westphal, Luderitz, and Bester *(2)*. This method relies on the following basic properties of lipopolysaccharide: the solubility of proteins, but not LPS, in phenol; the solubiliity of LPS in an aqueous environment (water); the total miscibility of phenol and water at elevated temperatures about 68°; and the relative ease by which phenol and water can be separated upon cooling and centrifugation. In general, this method is relatively uncomplicated and can be carried out even by investigators who are not generally accustomed to doing chemical extraction procedures. In general, the basic procedure involves Gram-negative bacteria that are disrupted in homogeneous solutions of equal volumes of phenol and water. When cooled to 5–10°C, the mixture resolves into three phases, an upper water layer (containing the LPS), a phenol layer, and at the interface between the two a variably sized layer of material that is both water and phenol insoluble. Extraction of the LPS into the upper water layer, that is then simply removed and subsequently manipulated, constitutes the essence of this method.

1. A 68°C water bath may be conveniently set up by placing a glass tray or 2-L beaker, filled halfway with water, on a hot plate/stirrer.
2. To make a solution of 90% phenol (w/v), add 10.8 g of crystalline phenol (if possible use freshly purchased and newly opened bottle) to a 50-mL graduated cylinder that contains a stir bar. Place the graduated cylinder in the 68°C water bath and add approx 1.2 mL of double-distilled H_2O (prewarmed to 68°C) to bring the volume to 12 ml. Stir briefly to dissolve the phenol (the crystals that will by themselves liquify at the 68°C temperature). This solution of phenol should be colorless. If it manifests any sign of discoloration, the phenol may be old and not ideal for extraction of LPS. Maintain the 90% phenol at 68°C until ready for use.
3. In a separate 50-mL beaker, suspend 2–4 g of wet Gram-negative bacteria (grown in standard fashion as described in the previous section) in 10 mL of double-distilled water and warm to 68°C. Stir the bacterial suspension at a moderate pace using a magnetic stir bar until a uniform paste white suspension is obtained.
4. Allow approx 10–15 min for the phenol and bacterial suspension to come to equilibrium at 68°C, at which time the 90% phenol should be added to the bacterial suspension in a 1:1 (volume:volume) ratio. Using a glass pipet, add 5 mL of the 90% phenol reagent drop-wise with constant stirring to the bacterial suspension. (It is sometimes helpful to pipet the 68°C water from the water bath into the glass pipet to heat the pipet glass to an elevated temperature.) The remaining balance of 5 mL may be added to the bacterial suspension more quickly. Mix continuously at 68°C for approx 10–20 min.
5. Transfer the suspension to a glass centrifuge tube on an ice bath and cool to 4°C. Centrifuge the mixture at 1800*g* for 25 min at 4°C. A clear to opalescent aqueous layer (sometimes with a yellowish or bluish tint, the "Tyndall" effect) will form on top. Below this will be an interphase of white-gray insoluble material that, depending upon the type of centrifuge used, may present as packed material with a 45° angle inclination. At the bottom of the tube is a bright golden layer of phenol containing primarily protein and usually accompanied by a relatively solid white or gray pellet of bacterial cell residue.
6. Using a pipet, very carefully remove as much of the aqueous layer as possible, being careful to disturb the integrity of the gray-white interface material as little as possible, keeping track of the total amount of aqueous phase removed. Pipet this into a glass centrifuge tube and maintain at 4°C.
7. Transfer all of the residual material (interface, phenol phase, and pellet) back to the glass extraction beaker and rinse the glass centrifuge tube (via vortexing) with a volume of double-distilled water exactly equal to that which was removed. Transfer this to the extraction beaker and reheat the entire mixture with continu-

ous mixing to 68°C for an additional 15 min. Repeat the centrifugation steps described above to generate a second aqueous extraction phase. Combine the aqueous layers.

8. Dialyze these aqueous phases extensively against double-distilled H_2O at 4°C until the residual phenol in the aqueous phase is totally eliminated. Use dialysis tubing with a MW cutoff of between 12,000 and 14,000. The speed with which this is accomplished depends on the volume of dialysate and the frequency with which the distilled water reservoir is changed (minimum time is approx 24 h). The absence of residual phenol is best and most sensitively monitored by sniffing the dialysis tube for the odor of phenol.
9. The major contaminant is usually nucleic acid (and primarily RNA). Removal of nucleic acids can be accomplished by digestion with RNase (40 µg/mL) and DNase (20 µg/mL) in the presence of 1 µL/mL of 20% $MgSO_4$ and 4 µL/mL of chloroform. Incubate at 37°C overnight. Dialyze once against 0.1 *M* acetate buffer (pH 5.0) and then against double-distilled H_2O three times.
10. Following treatment with nucleases, and in preparation for digestion of proteins, the suspension is made up to 0.01 *M* Tris at pH 8.0 by adding one-ninth the suspension volume of the LPS preparation of a stock solution of 0.1 *M* Tris, pH 8.0. Proteinase K is then added to a final concentration of 20 µg/mL. The suspension is heated in a water bath at 60°C for 1 h, and then overnight at 37°C. The suspension is then dialyzed once again against double-distilled H_2O for five to six exchanges, and finally lyophilized. The anticipated yields are between 20 and 50 mg of LPS with <2% protein and usually <1% nucleic acids (*see* **Note 3**).

3.3. Extraction of Bacterial Lipopolysaccharide: Phenol-Chloroform-Petroleum Ether

In general, the phenol-chloroform-petroleum (PCP) method is applicable for LPS extraction from a few grams to several hundreds of grams of dried bacteria *(6)*. The yield of extracted LPS, however, could be decreased significantly if a small initial amount of dried bacteria is used. Therefore, it is highly recommended to start LPS isolation with at least 5–10 g of bacteria. The following protocol was adopted for LPS extraction for 10 g of dried bacteria, and, therefore, can easily be scaled to meet the needs of individual investigators.

1. To prepare the extraction mixture, dissolve 90 g of crystallized phenol in 11–12 mL of deionized water and, then combine with chloroform and light petroleum ether in a volume ratio of 1:5:8 (*see* **Note 4**).
2. Add 40 mL of the extraction mixture to 10 g of the dried bacteria in a glass centrifuge tube. Maintaining the tube on ice, disperse bacteria in the extraction mixture by homogenizing with a medium-size rotor-stator generator (Ultra-Turrax laboratory homogenizer) until a fine bacterial suspension is obtained (*see* **Note 5**). If the resultant suspension is still very dense, add an additional 5–10 mL of the extraction mixture.

3. Extract LPS into the organic extraction solution at room temperature for 5–10 min.
4. Centrifuge the bacteria at 9000*g* for 15 min, and collect and save the supernatant, which should be a golden color above a white to brownish-white relatively well-packed pellet.
5. Repeat the extraction procedure with the remaining bacterial pellet by exactly following the steps as described above.
6. Combine the supernatants from the first and second extraction and filter them through a paper filter (Whatman, grade no. 3 filter paper) into a round-bottom flask that attaches via a ground glass fitting to a standard rotary evaporator distillation instrument.
7. Evaporate the petroleum ether and chloroform at 30–40°C on the rotary evaporator (R-114 Series, Brinkmann Instruments, Westbury, NY) under reduced pressure until only the crystallized phenol is remaining.
8. Add a minimal but sufficient amount of deionized water to dissolve the crystallized phenol.
9. Measure the resultant volume of phenol/LPS solution using a glass cylinder and transfer this into a centrifuge tube. Very slowly (drop-wise) add five volumes of diethyl ether-acetone to one volume of phenol/LPS (1:5, v/v) during constant stirring of the mixture on a magnetic stirrer, until precipitation of the flocculent white LPS from the phenol phase occurs. (You may add up to six volumes of diethyl ether:acetone.) If a precipitate has not been observed, allow the mixture to incubate at room temperature for 3 h to allow LPS precipitation.
10. Separate the precipitated LPS by centrifugation at 9000*g* for 15 min. Discard the supernatant and save the white pellet material (LPS).
11. Wash the extracted LPS once with 50 mL of 80% aqueous phenol (w/v) and three times with diethyl ether to remove residual traces of proteins and phenol respectively.
12. Dry LPS under the hood until the residual ether smell disappears.
13. To reduce contamination with bacterial RNA and DNA, dissolve the LPS in deionized water, disaggregate on ultrasonic bath (Fisher Ultrasonic Cleaner) for 5 min and then centrifuge at 100,000*g* for 4 h. Discard the supernatant and dissolve the sedimented LPS in deionized water. Dialyze LPS against deionized water for 3 d at 4°C (*see* **Note 6**) and then lyophilize.
14. LPS can finally be reconstituted in sterile deionized water at a concentration of 1 mg/mL (*see* **Note 7**) and stored at 4°C for several months in a tightly sealed tube provided that on each occasion prior to use, LPS is treated for 3 min on ultrasonic bath (*see* **Notes 8** and **9**).

4. Notes

1. Phenol is a carcinogen and is absorbed rapidly through the skin. Therefore, gloves should be worn when working with it. A fumehood should be used when heating phenol.

2. All cultures should be checked for contamination prior to and at the end of the growth cycle by culturing bacteria on LB agar (Difco) plates and controlling the shape of colonies formed.
3. Because polysaccharides are soluble in water and would not be removed by the methods described above, they may cause variable degrees of contamination of the LPS preparation. It is especially important to be aware of this when dealing with organisms that are likely to be encapsulated, such as *Klebsiella pneumoniae*, as the capsular polysaccharide may be extracted along the LPS.
4. If the resultant mixture is not a monophasic transparent solution, this would indicate the presence of water in the crystallized phenol. In such a case, add fractionally more solid phenol until the extraction mixture is clear.
5. Dispersion of bacterial suspension by homogenizer with rotor-stator generator does not break down the bacteria, but rather results in formation of a single-cell suspension and, therefore, this step increases the yields of PCP-extracted LPS.
6. For several days of dialysis, always use cold room conditions to eliminate the potential bacterial contamination of the sample.
7. To prepare a stock LPS solution, use chemically resistant borosilicate glass tubes that have reduced electrostatics as compared to plastic polypropylene tubes and, thereby, allow an easy introduction of LPS powder onto a tube. Always use lyophilized LPS that has been dried overnight in a vacuum over phosphorus pentoxide (cat. no. P0679, Sigma, St. Louis, MO), as LPS can absorb substantial amounts of moisture during storage. For this purpose, transfer an appropriate amount of lyophilized LPS into a glass borosilicate tube, the weight of which has been analytically measured and recorded. Place the tube with LPS in dessicator with phosphorus pentoxide and dry overnight under vacuum. On the following day, immediately measure the weight of the tube with LPS after the vacuum dessicator is opened. The amount of dried LPS is determined as the difference between the weight of the tube with dried LPS minus the weight of the empty tube.
8. The proximal portion of LPS possesses a number of negatively charged groups including phosphoryl groups of lipid A and the core, as well as carboxyl residues of 2-keto-3-deoxyoctonic acid (Kdo). Although the chemical structure of LPS suggests a strong repulsion between the molecules, in fact, the anionic properties of LPS are counterbalanced by the presence of both inorganic cations, such as Na^+, K^+, Mg^{2+}, and organic polyamines. The presence of neutralizing cations and polyamines drastically reduce the solubility of extracted LPS and, specifically, those of R-chemotypes because of their predominant hydrophobic properties associated with lipid A and augmented by the lack of polysaccharide tail. Apparently the bridging effect of divalent cations (Mg^{2+} and Ca^{2+}) play the key role in the aggregation state of the extracted LPS in aqueous solutions. To improve the solubility of LPS in aqueous solution, electrodialysis of LPS following their conversion into a triethylamine salt form was developed *(3)*. The dissociation activity of triethylamine seems to be associated with the bulky size of this

compound that prevents the tight binding of LPS molecules to each other yet not affecting the endotoxic properties of LPS. However, PCP-extracted and lyophilized R-LPS can be solubilized easily in sterile deionized water at a concentration of 1 mg/mL. To improve LPS solubility by its partial conversion into a triethylamine salt form, add directly 5 μL of triethylamine (cat. no. T 0886, Sigma) per one milliliter of LPS stock solution at 1 mg/mL. Check the pH of the LPS solution by placing a drop of it on an Alkacid Test Paper (Fisher) and adjust pH, if necessary.

9. Although the PCP method was developed originally for primary extraction of R-LPS, it can also be used for further purification of S-LPS that has first been extracted by a phenol-water procedure. The combination of these two extraction methods have the added advantage of a high yield of S-LPS achieved by LPS extraction into a hydrophilic aqueous phase (phenol-water extraction) and further S-LPS refining by a PCP re-extraction that removes such contaminants as RNA, DNA, proteins, and polysaccharides. Thus, the combination of two extraction procedures has been shown to very efficient in purification of *Bacteroides fragilis* LPS from contaminating capsular polysaccharides and glycan.
10. It is anticipated that using one or the other (or both) of the extraction and purification methods described in the preceding sections, virtually 98% of the investigative needs of most LPS researchers should be met. Because of this, attention in this chapter has not focused on a description of any of the other available methodologies. For example, there is a well-described butanol extraction procedure that some of the coauthors of this chapter have published ***(13)*** that results in the preparation of LPS in association with outer-membrane microbial proteins. Furthermore, a relatively rapid EDTA extraction of up to 50% of available LPS from bacteria has been described ***(14)***. Whereas both of these are useful techniques, they do not add substantially to the overall general utility of the two methods that have been described in detail. As a consequence, unless there are very compelling arguments against the use of the hot phenol-water procedure or the phenol-chloroform-petroleum-ether method, it is the opinion of the authors that one of these methods should be employed in any initial efforts to purify LPS/endotoxin from Gram-negative bacteria.

References

1. Boivin, A., Mesrobeanu, I., and Mesrobeau, L. (1933) Preparation of the specific polysaccharides of bacteria. *C. R. S. Soc. de Biol.* **113,** 490–492.
2. Westphal, O., Luderitz, O, and Bister, F. (1952) Uber die Extraktion von Bacterien mit Phenol-Wasser. *Z. Naturforsch.* **78,** 148–155.
3. Westphal, O. and Luderitz, O. (1954) Chemische erforschung von lipopolysacchariden gram-neagtiver bacterien. *Agnew Chemie* **66,** 407–417.
4. Morrison, D. C., Silverstein R., Lei, M.-G., Chen, T.-Y., and Flebbe, L. M. (1992) Bacterial endotoxin-structure function and mechanism of action, in *Natural Toxins: Toxicity, Chemistry and Safety* (Keeler, R. F., Mandava, N. B., and Tu, A. T., eds.), Alaken, Inc., Fort Collins, CO, pp. 301–315.

5. Hitchcock, P. J., Leive, L., Maleka, P. J., Rietschel, E. Th., Strittmatter, W., and Morrison, D. C. (1986) A review of lipopolysaccharide nomenclature: past, present and future. *J. Bacteriol.* **166,** 699–705.
6. Galanos, C., Luderitz, D., and Westphal, O. (1969) A new method for the extraction of R. lipopolysaccharide. *Eur. J. Biochem.* **9,** 945–949.
7. Hitchcock, P. J. and Brown, T. M. (1983) Morphological heterogencity among Salmonella lipopolysaccharide chemotypes in silver-stained polyacrylamide gels. *J. Bacteriol.* **154,** 269–277.
8. Tsai, C. M. and Frasch, C. E. (1982) A sensitive silver stain for detecting lipopolysaccharides in polyacrylamide gels. *Analyt. Biochem.* **119,** 115–119.
9. Tesh, V. L. and Morrison, D. C. (1988) The interaction of *E. coli* with normal human serum: factors affecting the capacity of serum to mediate lipopolysaccharide release. *Microb. Pathogen.* **4,** 175–187.
10. Hitchcock, P. J. and Morrison, D. C. (1984) The protein component of bacterial endotoxin, in *Handbook of Endotoxin, Chemistry of Endotoxin*, vol. 1 (Rietschel, E. Th., ed.), Elsevier Science Publishers, Amsterdam, The Netherlands, pp. 339–375.
11. Skidmore, B. J., Morrison, D. C., Chiller, J. M., and Morrison, D. C. (1975) Immunologic properties of bacterial lipopolysaccharide. II. The unresponsiveness of C3H/HeJ mouse splenocytes to LPS-induced mitogenesis is dependent upon the method used to extract LPS. *J. Exp. Med.* **142,** 1488–1508.
12. Morrison, D. C., Betz, S. J., and Jacobs, D. M. (1976) Isolation of a lipid A-bound polypeptide responsible for "LPS-initiated" mitogenesis of C3H/HeJ spleen cells. *J. Exp. Med.* **144,** 840–846.
13. Morrison, D. C. and Leive, L. (1975) Fractions of lipopolysaccharide from E. coli O111:B4 prepared by two extraction procedures. *J. Biol. Chem.* **250,** 2911–2919.
14. Leive, L. and Morrison, D. C. (1972) Isolation of lipopolysaccharide from bacteria, in *Methods in Enzymology, Complex Carbohydrates*, Vol. XXVII (Ginsberg, V., ed.), Academic Press, New York, pp. 254–262.

3

Assay of Anti-Endotoxin Antibodies

Lore Brade

1. Introduction

Lipopolysaccharides (LPS) constitute components of the outer membrane of Gram-negative bacteria. Chemically, they consist of a heteropolysaccharide and a covalently linked lipid, termed lipid A. The polysaccharide region is made up of the O-specific chain (built from repeating units of three to eight sugars) and the core part, divided into the inner core (the part linked to the lipid) and the outer core (the part linked to the O-specific chain). LPSs possessing an O-specific chain are called smooth LPS (S-LPS), those not having an O-chain are termed rough (R-LPS). The latter type of LPS may be observed in mutants that have lost the ability to synthesize the O-chain, or in wild-type bacteria without known genetic defect. LPS also represent the endotoxin of Gram-negative bacteria. In mammals, including humans, LPS exhibits a variety of biological effects that may be beneficial if administered in low amounts but harmful when present in higher concentrations as in the case of Gram-negative infection and Gram-negative septicemia.

Because of its surface exposure, LPS is a strong immunogen, inducing the formation of antibodies after experimental or natural infection or after experimental hyperimmunization. Antibodies against LPS are useful for the determination of different serotypes within a given bacterial genus and are used routinely in clinical and diagnostic laboratories. Especially for epidemiological surveys, taxonomic determination at the serotype level is of diagnostic value to follow outbreaks of epidemics. Most of the antibodies used for this purpose are directed against the O-specific chain of S-LPS, as it occurs in many pathogenic Gram-negative bacteria including *Enterobacteriaceae*, *Vibrionaceae*, *Pseudomonadaceae*, *Brucellaceae*, *Legionella*, and *Campylobacter*. As these

From: *Methods in Molecular Medicine, Vol. 36: Septic Shock*
Edited by: T. J. Evans © Humana Press Inc., Totowa, NJ

structures are characteristic for a given bacterium, the diagnostic power of the corresponding antibodies depends mainly on their epitope specificity, which is best met by monoclonal antibodies prepared against structurally defined O-antigens. On the other hand, the determination of LPS antibodies in patient sera or other body fluids may be used to diagnose an infection with a given bacterium, the isolation and identification of which could not be achieved during the acute infection. The use of defined LPS antigens is a prerequisite to get unequivocal results *(1)*.

Besides the O-antigenic determinants, other carbohydrate epitopes are located in the LPS molecule residing in the core region or in the lipid moiety or both. The core epitopes may be cryptic or accessible as antigens in S-LPS, but lipid A immunogenicity and antigenicity is never exposed in native S- or R-LPS *(2)*. As cleavage of the core-lipid A linkage is required to set free lipid A immunoreactivity, lipid A represents a neoantigen *(13)*. Lipid A antibodies are useful experimental tools for the detection and quantification of lipid A, because epitopes present in lipid A of all LPS studied so far are structurally related *(3–7)*.

Finally, antibodies against LPS are potential therapeutic agents in the fight against lethal endotoxic shock. As the elimination of endotoxin from the circulation during Gram-negative sepsis interferes with the very early events of the cascade leading ultimately to multiorgan failure and death, antibody therapy may be a very effective immunotherapy for this life-threatening disease.

For the reasons mentioned above, antibodies against the O-specific chain or lipid A cannot be used in therapy. However, those antibodies recognizing epitopes of the core region independent of the presence or absence of an O-chain may recognize a large spectrum of Gram-negative bacteria. We have developed such an antibody interacting with LPS of all serotypes of *Salmonella enterica*, *Escherichia coli*, and *Shigella*. In addition to being cross-reactive among the different strains in vitro, the antibody is cross-protective in vitro and in vivo in endotoxin and infection models *(8,9)*.

It is thus evident that LPS and lipid A serology provide many clinical and experimental applications. Reliable LPS serology requires defined, preferably monoclonal, antibodies and structurally characterized LPS antigens. Most importantly, however, LPS serology requires experience working with amphipathic molecules, which pose many more problems to an investigator than protein antigens. The following description may help newcomers in the field avoid having to learn by trial and error.

2. Materials

1. Polyvinyl chloride microtiter plates (96-well) flexible, U-shape (*see* **Note 1**) (Falcon, Los Angeles, CA, Becton Dickinson, cat. no. 3911).

2. Antigens: LPS and lipid A from any LPS, preferably from *E. coli* Re mutant strain F515 (*see* **Note 2**), kept frozen at a concentration of 1 mg/mL in water.
3. Micropipets (1–100 µL and 10–1000 µL) and disposable tips (Eppendorf, Germany).
4. Multichannel pipeter and tips (Eppendorf).
5. Phosphate-buffered saline (PBS): 136.9 m*M* NaCl, 2.68 m*M* KCl, 8 m*M* Na_2HPO_4, 2.4 m*M* KH_2PO_4, pH 7.2. Add 1% thimerosal (highly toxic!) to 0.01% final concentration. Make up in deionized distilled water, store at room temperature, use for 1 wk only.
6. Blocking buffer, PBS-casein (PBS-C): dissolve 25 g casein powder (Sigma, St. Louis, MO, purified powder from bovine milk, C-5890) in 0.3 *M* NaOH (800 mL) by overnight stirring at 37°C, cool down to room temperature, titrate to approx pH 7.5 with HCl (25%), add KH_2PO_4 and Na_2HPO_4, 10 m*M* each, and titrate finally to pH 7.2. Add thimerosal (1% stock solution) to 0.01% final concentration. Fill up to 1 L with deioized water. Store at –20°C in aliquots, do not freeze and thaw more than three times.
7. First antibody: human or animal sera (in general rabbit and mouse), monoclonal antibodies (MAbs).
8. Second antibody: horseradish peroxidase (HRP)-labeled conjugated anti-immunoglobulin (from Jackson Immuno Research, West Grove, PA) either goat anti-human IgG, IgA, or IgM (γ, α, and µ-chain specific, respectively) or goat anti-rabbit or goat anti-mouse IgG (heavy and light chains). All antibody preparations are usually used at a dilution of 1 : 1000 to 1 : 2000.
9. Substrate buffer: 0.1 *M* Na-citrate, adjust pH with citric acid to pH 4.5. Stable for 4 wk at 4°C.
10. Substrate solution: dissolve 2,2′-azino-di-3-ethylbenzthiazoline-6-sulfonic acid (ABTS) (irritant!) in substrate buffer (1 mg/mL) with sonication in an ultrasound water bath for 1 min, then add hydrogen peroxide (25 µL of a 0.1% stock solution in water). This solution should be prepared immediately before use. The 0.1% H_2O_2 stock solution should be stored at 4°C in a brown flask for not longer than 1 wk.
11. Stopping reagent: 2 % aqueous oxalic acid (harmful!). Stable for 6 wk at 4°C.
12. Microtiter plate reader (Dynatech MR 5000) at 405 nm (reference filter, 490 nm).
13. Optional equipment: nonautomated plate-washer from Nunc (Wiesbaden-Biebrich, Germany): Immuno Wash 12 (autoclavable), three-dimensional rocking table (Heidolph, Kelheim, Germany, poly-max 1040), peroxidase (POD) substrate enhancer (Boehringer Mannheim, Mannheim, Germany).

3. Methods

1. Prepare a solution of antigen in PBS to coat plates (*see* **Note 2**). The optimal concentration of antigen is determined empirically, but the usual range is 1–4 µg/mL for lipid A and 2–8 µg/mL for LPS (*see* **Note 3**). PBS and PBS-containing solutions are supplemented with thimerosal to a final concentration of 0.01%.

2. Add 50 µL of the antigen solution to each well of the 96-well plate. Cover the plate and incubate overnight at 4°C.
3. Flick out the antigen solution. Wash the plate four times with PBS. Use for washing a polyethylene bottle with a wide-necked tube for rinsing each well, but care must be taken to obtain a similar pressure over all the wells on the plate. Rinse each well with PBS, avoiding airbubbles. Flick out the washing buffer over a sink. Rinse the plate and flick out the fluid four times in total. Finally flick out the washing buffer, then rap the plate on a wad of paper towels until the towels show no more fluid. It is essential not to leave residual washing buffer in any well, since this will dilute the subsequent reagent (*see* **Note 4**).
4. Add 200 µL blocking buffer (PBS-C) (*see* **Subheading 2., item 6**) to all wells and incubate 1 h at 37°C, flick out the blocking buffer, then rap the plate on a wad of paper towels until the towels show no fluid (*see* **Note 5**).
5. For antibody determinations, make an appropriate dilution of the test serum in PBS-C (*see* **Note 6**). Add 50 µL of PBS-C to each well of two rows (we recommend to test each serum in duplicate) of the antigen-coated microtiter plate except the first two wells. To these wells add 100 µL each of the prediluted test serum. Transfer with a multichannel pipeter 50 µL from these two wells to the next wells, mix, and transfer again to the next wells and so on, mixing the contents of each well a minimum of six times before the following transfer. Cover the plate, put it into a wet chamber on a three-dimensional rocking table (optional) and incubate 1 h at 37°C.
6. Wash the plate four times with PBS as in **step 3**, above.
7. Dilute the peroxidase anti-immunoglobulin conjugate appropriately in PBC-C (*see* **Note 7**). Add 50 µL of conjugate to each well and incubate 1 h at 37°C.
8. Wash the plate four times with PBS as in **step 3** and then additionally two times with substrate buffer.
9. Add 50 µL substrate solution (*see* **Note 8**) to each well and incubate 30 min at 37°C. Stop the reaction by the addition of 2% aqueous oxalic acid (*see* **Note 9**).
10. Plates are read with a microtiter plate reader at 405 nm and titers are determined as the highest dilution of antiserum yielding an optical density of >0.2 at 405 nm.

4. Notes

1. Polyvinyl chloride is especially suited for the immobilization of amphiphathic antigens such as LPS or lipid A because of its hydrophobic properties. Never use Tween or any other detergent in the test. Note the batch-number of the plates; slight batch-to-batch variations are possible.
2. Soluble antigens such as lipid A or LPS can be adsorbed passively onto the polyvinyl plate. A prerequisite for an efficient coating is that the lipid A or LPS stock solution (do not use concentrations > 1–2 mg/mL in distilled water) is a clear or at least opalescent solution *(10)*. In a pure thermodynamical sense the term "solution" is not correct. Actually these are suspensions which, however, look like true solutions when well prepared. As an indicator for sufficient solubilization one can transfer an aliquot into an Eppendorf tube and centrifuge it for 10 min at

maximal speed in an Eppendorf centrifuge. The absence of any sediment indicates that your sample is well solubilized. Turbid solutions indicate the presence of larger aggregates. With such aggregates reproducible coating of plates is not possible, especially because these aggregates tend to detach from the plate. The solubility of lipid A is greatly influenced by its content of phosphate, e.g., bisphosphorylated lipid A is more water soluble than the monophosphorylated partial structure. The former is best suited as a screening antigen for lipid A antibodies, as all lipid A antibody specificities known so far react with bisphosphorylated lipid A ***(11)***.

3. The amount of lipid A antigen needed for efficient coating of plates is greatly influenced by the hydrophobicity of the compound. Using a series of synthetic lipid A preparations that possess the same hydrophilic backbone but a different acylation pattern, it was demonstrated that the varying degrees of hydrophobicity indeed influence the coating behavior of the lipid A-based compounds but not their specificity ***(11,12)***: A high number of fatty acids (4–7) yielded the highest reactivity. The coating efficiency of compounds containing four and seven fatty acids is not significantly different but for a compound with only two fatty acids, the amount for coating needed to get the same reactivity as with the more hydrophobic ones can be up to 30-fold higher ***(12)***. As a standard amount for coating of lipid A, we recommend a range between 1–4 μg/mL when polyclonal antisera are tested.

   For the characterization of monoclonal antibodies, however, more information is obtained from checkerboard titrations, in which both antigen and antibody are serially diluted. The antigen dilutions should cover a range from 4 to 0.032 μg/mL for lipid A and 10 to 0.08 μg/mL for LPS. From such checkerboard titrations enough data are obtained to set up binding curves that give valuable informations about the affinity of the antigen-antibody complex. In **Figs. 1** and **2** examples of checkerboard titrations are given. In **Fig. 1**, the binding of two different lipid A MAbs ***(13)***, in **Fig. 2** a MAb against the LPS of *Klebsiella pneumoniae* is shown ***(14)***. In **Fig. 1A**, an example of excellent binding over all tested antigen concentrations is illustrated, whereas **Fig. 1B** illustrates the behavior of a MAb with lower affinity. With decreasing amounts of antigen, the antibody concentration is clearly increasing, but all binding curves show similar slopes. In **Fig. 1B**, however, the steepness of the binding curves decreases with lower antigen concentration, indicating a low affinity of this MAb. When numerous antigens have to be tested with several different MAbs, useful information can be obtained already by testing only two or three concentrations of antigen, covering a broad range, e.g. 4, 0.4, and 0.04 μg/mL.

   Another important point for the coating of LPS on enzyme immunoassay (EIA)-plates concerns the nature of LPS, e.g. whether it is LPS of the R- or S-type. Usually all R-LPS coat EIA-plates efficiently at approx 4 μg/mL. In the case of S-LPS, however, higher amounts of antigen (10 μg/mL) are needed for coating because of its higher hydrophilicity depending on the number of repeating units of the O-specific chain. As the capacity of an EIA well is limited ***(13)***,

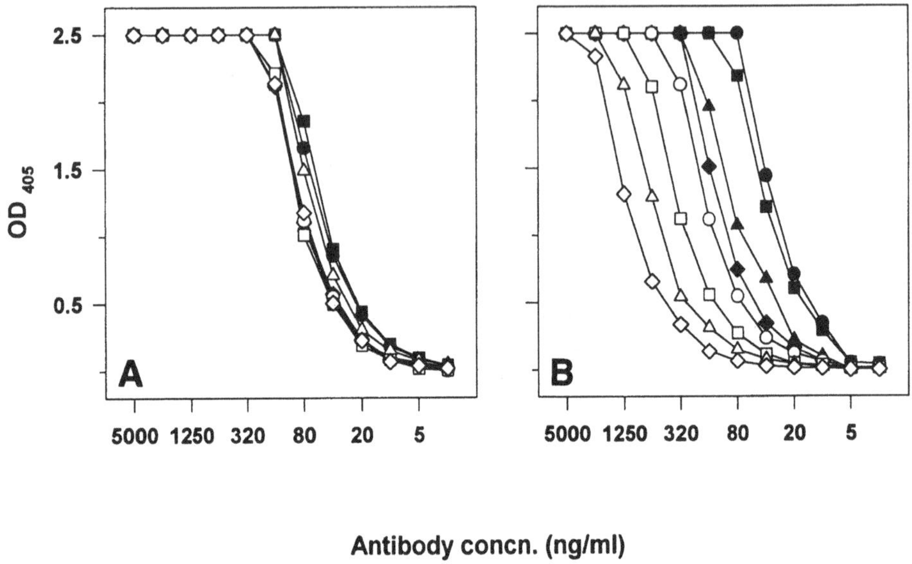

Fig. 1. Checkerboard titrations of two lipid A MAbs (A6, **A** and S 1–15, **B**) in EIA using as solid-phase antigen synthetic tetracyl lipid A ***(11,13)***. Plates were coated with graded concentrations of antigen corresponding to 4 (filled circles), 2 (filled squares), 1 (filled triangles), 0.5 (filled diamonds), 0.25 (open circles), 0.125 (open squares), 0.063 (open triangles), and 0.032 (open diamonds) µg/mL using 50 µL/ well. MAbs A6 and S 1–15 were added at the concentrations indicated on the abscissa. Values are the means of quadruplicates (confidence values do not exceed 10%).

> 500 ng/well should not be used because the excess antigen may be bound loosely and detach later together with bound antibody.

4. In solid-phase assays, washing steps are required to remove unbound reagents such as antigen, first, or second antibody. Thus, the washing steps are most crucial in EIA to achieve reproducibility. One should keep in mind that gentle washes favor the detection of low-affinity binding ***(15)***. Usually, we prefer intensive washing steps because of an optimal signal-to-noise ratio and much better reproducibility, but when low-affinity binding is of particular interest, gentle washing protocols may be useful for their detection. Automatic EIA-plate washers are commercially available and, theoretically, they should give a better standardization of the washing steps. However, special care must be taken to maintain the instrument. In our laboratory, the most reproducible results are obtained applying hand-washing by an experienced person.
5. The sensitivity of the EIA test system implies a stringent limit on the acceptable background signal because of nonspecifically bound reactants. Low background is usually achieved by thorough blocking of the wells with nonspecific serum, BSA, skimmed dried milk, gelatin, or detergents. One of the most effective blocking reagents is the nonionic detergent Tween-20. Unlike other reports,

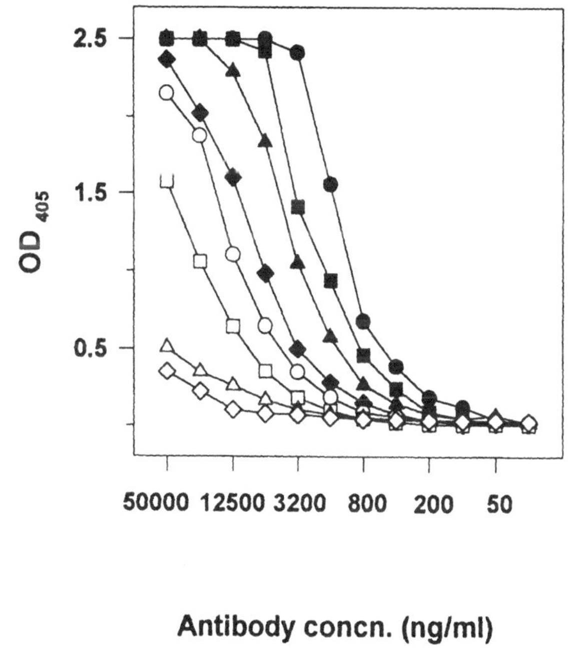

Fig. 2. Checkerboard titration of a MAb (S 47–19) against the LPS of *Klebsiella pneumoniae* in EIA using as solid-phase antigen LPS of *K. pneumoniae* R 20 ***(14)***. Plates were coated with graded concentrations of antigen corresponding to 10 (filled circles), 5 (filled squares), 2.5 (filled triangles), 1.25 (filled diamonds), 0.63 (open circles), 0.32 (open squares), 0.16 (open triangles), and 0.08 (open diamonds) μg/mL using 50 μL/ well. The MAb concentrations are indicated on the abscissa. The confidence limits of quadruplicate samples did not exceed 10%.

we found that lipid A and LPS are detached to a large extent from polyvinyl-plates by detergent-containing buffers ***(12)***. Thus, detergents have to be replaced by a highly effective blocking reagent. We found casein to be the most effective blocking reagent. Thus, omitting detergents and using casein as blocking reagent results in a good reproducibility and high sensitivity with low amounts of antigen.

6. Although blocking with casein is an efficient step, human and animal sera are also able to produce—in a certain range of dilution—a positive signal on control plates without antigen. This reactivity is called background reactivity. The control plate has to be processed in parallel with the test plate. The coating procedure is also done on the control plate with PBS without antigen. From our experience, most human and rabbit sera in the range of 1:500–1000 and mice sera in the range of 1:200–500 exhibit no more background reactivity. MAbs should also be tested for background reactivity.
7. The bound first antibody (serum or MAb) is usually detected with a second, polyclonal antibody directed against the constant region of the immunoglobulin

of the corresponding species. Covalently linked to this second antibody is an enzyme that reacts with a chromogenic substrate. We use HRP-conjugated antisera, as HRP has a wide range of substrates and the conjugate is cheaper than the alkaline phosphatase-conjugate. Before use, we titrate any new batch of second antibody in a range of 1:500 to 1:8000. In most cases, no clear plateau can be observed, but there should exist a range of at least two titer steps in which the OD values are comparable (not more than a difference of approx 0.3 in an OD range >1.0). With the second antibody (from Jackson Immuno Research) the range is typically between 1:1000–2000. Second antibodies that can be used only at dilutions of ≤ 1:500 should not be used as they produce high background reactivity. Be cautious that the second antibody raised in rabbits does not contain specific natural antibodies against the LPS under investigation. We have observed natural antibodies against *Bordetella pertussis* and *Acinetobacter spp.* quite frequently.

8. As substrate for the HRP-conjugated second antibody we prefer the substrate ABTS for several reasons:
   a. ABTS is not carcinogenic like other HRP-substrates.
   b. The color development is not too fast (optimal conditions are 30 min at 37°C), so that the time period can be easily taken into account, when many plates have to be handled at a time.
   c. In a geometric dilution row with, i.e., a factor of two of the first antibody, the color intensity correlates linearly with the concentration of antibody over a wide range.
   d. A substrate enhancer exists for ABTS, which results in an enhancement of an OD_{405} of approx 0.3–0.5 under the given test conditions. For supernatants of MAbs with low reactivity and affinity this reagent may be helpful.
9. The plates should be subjected to the EIA-reader immediately after adding the stop solution. With ongoing time the OD values decrease as the chromophore formed is not stable. Leaving the plates for 0.5 or 1 h at room temperature or at 4°C before reading results in a decrease of OD values by approx 5 or 10%, respectively.

References

1. Rietschel, E. T., Brade, H., Holst, O., Brade, L., Müller-Loennies, S., Mamat, U., Zähringer, U., Beckmann, F., Seydel, U., Brandenburg, K., Ulmer, A. J., Mattern, T., Heine, H., Schletter, J., Loppnow, H., Schönbeck, U., Flad, H.-D., Hauschildt, S., Schade, U. F., Di Padova, F., Kusumoto, S., and Schumann, R. R. (1996) Bacterial endotoxin: chemical constitution, biological recognition, host response, and immunological detoxification, in *Current Topics in Microbiology and Immunology, Pathology of Septic Shock* (Rietschel, E. T. and Wagner, H., eds.), Springer-Verlag, Berlin, Germany, pp. 39–81.
2. Galanos, C., Freudenberg, M. A., Jay, F., Nerkar, D., Veleva, K., Brade, H., and Strittmatter, W. (1984) Immunogenic properties of lipid A. *Rev. Infect. Dis.* **6,** 546–552.

3. Galanos, C., Lüderitz, O., and Westphal, O. (1971) Preparation and properties of antisera against the lipid A component of bacterial lipopolysaccharides. *Eur. J. Biochem.* **24,** 116–122.
4. Brade, L. and Brade, H. (1985) Characterization of two different antibody specificities recognizing distinct antigenic determinants in free lipid A of *Escherichia coli. Infect. Immun.* **48,** 776–781.
5. Brade, L., Rietschel, E. T., Kusumoto, S., Shiba, T., and Brade, H. (1986) Immunogenicity and antigenicity of synthetic *Escherichia coli* lipid A. *Infect. Immun.* **51,** 110–114.
6. Brade, L., Rietschel, E. T., Kusumoto, S., Shiba, T., and Brade, H. (1987) Immunogenicity and antigenicity of natural and synthetic *Escherichia coli* lipid A. *Prog. Clin. Biol. Res.* **231,** 75–97.
7. Brade, L., Brandenburg, K., Kuhn, H.-M., Kusumoto, S., Macher, I., Rietschel, E. T., and Brade, H. (1987) The immunogenicity and antigenicity of lipid A are influenced by its physicochemical state and environment. *Infect. Immun.* **55,** 2636–2644.
8. Di Padova, F. E., Brade, H., Barclay, R., Poxton, I. R., Liehl, E., Schuetze, E., Kocher, H. P., Ramsay, G., Schreier, M. H., McClelland, D. B. L., and Rietschel, E. T. (1993) A broadly cross-protective monoclonal antibody binding to *Escherichia coli* and *Salmonella* lipopolysaccharides. *Infect. Immun.* **61,** 3863–3872.
9. Bailat, S., Heumann, D., Le Roy, D., Baumgartner, J. D., Rietschel, E. T., Glauser, M. P., and Di Padova, F. (1997) Similarities and disparities between core-specific and O-side-chain-specific antilipopolysaccharide monoclonal antibodies in models of endotoxemia and bacteremia in mice. *Infect. Immun.* **65,** 811–814.
10. Galanos, C. and Lüderitz, O. (1975) Electrodialysis of lipopolysaccharides and their conversion to uniform salt forms. *Eur. J. Biochem.* **54,** 603–610.
11. Kuhn, H.-M., Brade, L., Appelmelk, B. J., Kusumoto, S., Rietschel, E. T., and Brade, H. (1992) Characterization of the epitope specificity of murine monoclonal antibodies directed against lipid A. *Infect. Immun.* **60,** 2201–2210.
12. Kuhn, H.-M., Brade, L., Appelmelk, B. J., Kusumoto, S., Rietschel, E. T., and Brade, H. (1993) The antibody reactivity of monoclonal lipid A antibodies is influenced by the acylation pattern of lipid A and the assay system employed. *Immunobiology* **189,** 457–471.
13. Brade, L., Engel, R., Christ, W. J., and Rietschel, E. T. (1997) A nonsubstituted primary hydroxyl group in position 6′ of free lipid A is required for binding of lipid A monoclonal antibodies. *Infect. Immun.* **65,** 3961–3965.
14. Süsskind, M., Brade, L., Brade, H., and Holst, O. (1997) Identification of a novel heptoglycan of α→2-linked D-*glycero*-D-*manno*-heptopyranose. Chemical and antigenic structure of lipopolysaccharides from *Klebsiella pneumoniae* ssp. *pneumoniae* rough strain R20 (01⁻:K20⁻). *J. Biol. Chem.* **273,** 7006–7017.
15. Kuhn, H.-M. (1993) Cross-reactivity of monoclonal antibodies and sera directed against lipid A and lipopolysaccharides. *Infection* **21,** 179–186.

4

Isolation of the Bactericidal/Permeability-Increasing Protein from Polymorphonuclear Leukocytes by Reversible Binding to Target Bacteria

Jerrold Weiss

1. Introduction

Polymorphonuclear leukocytes (PMN) play a prominent role in host defense in mammals against invading bacteria *(1,2)*. Among the essential attributes of these highly specialized cells are the elaboration of an array of cytotoxic peptides and polypeptides that can be targeted at bacterial prey. This includes the bactericidal/permeability-increasing protein (BPI), a cytotoxic protein that at n*M* concentrations acts selectively against many Gram-negative bacteria. The principal determinant of the target-cell selectivity and potency of BPI is its ability to bind avidly to lipopolysaccarides (LPS), abundant glycolipids found uniquely in the outer leaflet of the outer membrane of these bacteria *(3,4)*. Binding of BPI to bacterial outer membrane LPS not only initiates antibacterial cytotoxicity but also blocks the potent pro-inflammatory activity of this bacterial product *(5)*. Host responses to LPS fuel the delivery of host defenses (e.g., PMN) to sites of infection but can also lead to profound inflammatory injury if inadequately regulated *(4,6)*. By contributing to the eradication of the bacteria that produce LPS and by blocking the activity of LPS already present, BPI can play a major role in elimination of invading Gram-negative bacteria and in downregulating further host responses to these bacteria.

Studies of BPI were made possible initially by developing techniques for isolation of this highly cationic protein. Although methods for expression and purification of recombinant species of BPI are now available *(7)*, in most situations the purification of BPI from its native source (PMN) remains the most direct means to procure functionally significant amounts of this protein. This is

From: *Methods in Molecular Medicine, Vol. 36: Septic Shock*
Edited by: T. J. Evans © Humana Press Inc., Totowa, NJ

particularly true if PMN-rich inflammatory exudates can be induced in experimental animals as this provides a highly enriched and abundant starting material for extraction and isolation of BPI. Accordingly, in this chapter, methods for isolation of BPI from peripheral blood PMN of humans and from peritoneal exudate PMN from New Zealand white rabbits will be detailed. From both cellular sources, purification of BPI depends on release of the protein in soluble form from the PMN by harsh acid extraction ***(8,9)***. BPI in PMN acid extracts can be purified by a combination of ion-exchange and reverse-phase chromatographies. However, an even more efficient method has been developed that relies upon the avid and reversible binding of BPI to the surface of sensitive Gram-negative bacteria ***(10)***, as described in this chapter (*see* **Fig. 1**).

2. Materials

2.1. Collection of PMN from Human Peripheral Blood (*see* Note 1)

1. Peripheral blood from normal human volunteers.
2. Sterile syringes (60 mL), monoject, sterile, nonpyrogenic (Becton-Dickinson, Rutherford, NJ).
3. Sodium heparin.
4. Dextran T-500 (Pharmacia, Piscataway, NJ).

2.2. Collection of PMN from Sterile Rabbit Peritoneal Inflammatory Exudates

1. New Zealand white rabbits (2–3 kg; ≥ 6 mo).
2. Oyster glycogen.
3. Sodium chloride (0.9%) for irrigation (sterile, nonpyrogenic; Baxter, Deerfield, IL).
4. Needles (19-gage), winged infusion set (Terumo, Medical, Elkton, MD).
5. Disposable hypodermic needles (16-gage) (Becton-Dickinson).
6. Polypropylene, 50-mL sterile conical centrifuge tubes.

2.3. Acid Extraction of PMN

1. Potter-Elvehjem homogenizer and motor-driven Teflon-coated pestle (Koates, New Brunswick, NJ).

2.4. Affinity Adsorption of BPI to Target Bacteria

1. K-12-based strains of *Escherichia coli*.
2. Nutrient broth (0.8% w/v) dissolved in sterile physiological saline ± 10 m*M* phosphate buffer at pH 7.4.

2.5. Reversed-Phase High Performance Liquid Chromatography (HPLC)

1. C4 column (Vydac; The Separations Group, Hesperia, CA).

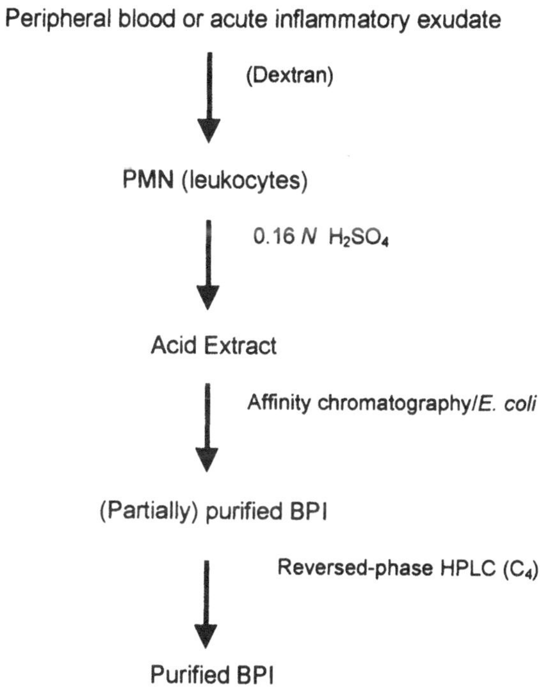

Fig. 1. Flow diagram of steps in affinity purification of BPI.

3. Methods

3.1. Collection of Human PMN

1. Collect peripheral blood from normal healthy volunteers into syringes containing sodium heparin (14 U/mL, final concentration) as anticoagulant.
2. Mix blood 1 : 1 (v/v) with 2% (w/v) Dextran T-500 in cylindrical-shaped glass or polypropylene vessel.
3. Sit at room temperature to allow sedimentation of red blood cells by gravity (approx 20–40 min).
4. Remove white-blood-cell-enriched upper phase (red-blood-cell-depleted) and collect leukocytes by sedimentation at 1000*g* for 10 min at room temperature.
5. Wash sedimented leukocytes once with physiological saline.
6. Contaminating red cells (*see* **Note 2**) can be lysed by incubation of cells in 10 vol of 0.87% (w/v) ammonium chloride at 37°C for 10 min; leukocytes are recovered by sedimentation at approx 100*g* for 5 min and washed twice with saline.
7. Store cell pellets at –70°C until use for BPI purification (*see* **Note 3**).

3.2. Collection of PMN from Sterile Rabbit Peritoneal Inflammatory Exudates

1. Freshly prepare a supersaturated solution of oyster glycogen (2.5 mg/mL) in sterile, pyrogen-free physiological saline.

2. Briefly incubate solution at 37°C.
3. Inject 250–300 mL of glycogen-saturated saline into the peritoneal cavity of a rabbit, using a 60-mL syringe that is simultaneously connected via a three-way stopcock into a bottle containing the injected solution and a winged infusion set to deliver intraperitoneally the glycogen-saline.
4. At 12–16 h after injection (*see* **Note 4**), the inflammatory peritoneal exudate is collected by insertion of a 16-gage needle into the peritoneal cavity. The rabbit is placed on a small wooden table with a hole placed in the middle to allow protrusion of the animal's abdomen. This permits access to the peritoneal cavity for insertion of the needle and collection of the exudate by gravity flow.
5. The exudate cells are separated from the inflammatory fluid by sedimentation of the cells at 100–200*g* for 5 min. The cells are washed once with saline and stored at –70°C until use.

3.3. Acid Extraction of PMN

1. Stored frozen cells are thawed briefly at room temperature and then placed in an ice-water bath.
2. Iced distilled water is added to thawed cell pellets to bring cell concentration to approx 3×10^8 PMN/mL.
3. Homogenize PMN suspension in ice bath using motor-driven homogenizer (approx 100 strokes/15–30 min; *see* **Note 5**).
4. Add 2 vol of ice-cold 0.4 N H_2SO_4 to 3 vol of PMN homogenate (final concentration of 0.16 N H_2SO_4).
5. Sit in ice-bath for 30 min with periodic mixing by vortexing (every 10 min for 5–10 sec each time).
6. Spin extract at 23,000*g* for 20 min at 4°C.
7. Collect 23,000*g* supernatant and dialyze either against 10 m*M* sodium acetate/acetic acid buffer at pH 4.0 or 2 m*M* Tris-HCl at pH 7.4 (*see* **Note 6**) at 4°C.
8. After sample is fully equilibrated with dialysis buffer, spin at 23,000*g* for 20 min at 4°C. Recovered supernatant can be stored at 4°C for several months.

3.4. Affinity Purification of BPI from PMN Acid Extract Using BPI-Sensitive Target Bacteria

1. Frozen stock cultures of *E. coli* (*see* **Note 7**) are inoculated (1:100, v/v) into nutrient broth (0.8% nutrient broth [w/v] in physiological saline) and incubated overnight at 37°C.
2. Overnight, stationary-phase, cultures are diluted 1:50 (v/v) into fresh medium and incubated at 37°C for approx 3 h until growing bacteria are in mid-late logarithmic phase (approx 5×10^8 bacteria/mL).
3. Bacteria are collected by sedimentation at approx 5000*g* for 5 min at room temperature, washed once with saline, and resuspended in buffered 0.9% (w/v) NaCl at pH from 4.0 to 7.5 (*see* **Note 8**) at suitable bacterial concentration (*see* **step 4**).
4. Bacteria and dialyzed PMN extract are mixed, typically at a bacterial concentration of 5×10^8 bacteria/mL (*see* **Note 9**), and incubated at 37°C for 15 min with shaking.

5. Unbound proteins are removed by sedimentation of bacteria (as above) and one wash of sedimented bacteria with 10 m*M* acetate-buffered medium.
6. Washed bacteria are resuspended in acetate buffer supplemented with 0.2 *M* $MgCl_2$ at 5×10^9 bacteria/mL and incubated for 10 min at 37°C (*see* **Note 10**).
7. Eluted proteins are recovered in the supernatant fraction following sedimentation of bacteria (*see* **step 3**).
8. An aliquot of the recovered supernatant fraction is analyzed by reversed-phase HPLC on an analytical Vydac C4 column to assess the purity and yield of eluted BPI (*see* **Note 11**). Proteins are eluted by using a linear gradient of acetonitrile (0–95%, v/v) in 0.1% trifluoroacetic acid developed over 30 min at a flow rate of 1 mL/min. If necessary, the procedure is repeated with the rest of the sample to achieve further purification of BPI.
9. Eluted BPI is promptly dialyzed against 10–20 m*M* acetate buffer at pH 4.0 at 4°C (*see* **Note 12**).
10. Recovered BPI, after, dialysis, is analyzed by analytical reversed-phase HPLC (*see* **step 8**, above) to ensure purity and assess yield of BPI.

4. Notes

1. PMN are the richest natural cellular source of BPI *(11)*. Rabbit and human PMN contain approx 0.5–1.0 mg of BPI/10^9 PMN *(11–13)*. Approximately 10^9 PMN can be recovered from 300–400 mL of normal blood. Leukaphoresis of patients with chronic myeloid leukemia can provide much larger numbers of BPI-containing cells (up to 10^{11}) for purification provided the cells are at or beyond the promyelocytic stage of differentiation *(11)*.
2. For purposes of purification of BPI, separation of PMN from other leukocytes is unnecessary. Thus, additional cellular purification steps should be avoided to maximize recovery of PMN. However, the presence of >5 red blood cells/leukocyte can yield contaminants in the acid extracts that complicate subsequent purification. Hence, if red-cell contamination exceeds this level after dextran sedimentation, the ammonium chloride lysis step should be included.
3. BPI is localized in the primary granules of PMN *(11,14)*. Granule-rich fractions can prepared by homogenization of PMN in osmotically protected medium (e.g., 0.34 *M* sucrose) to provide a more enriched starting material. However, recovery of granules in these fractions is generally ≤50% and requires working with fresh cells. Therefore, ultimate yields of purified BPI are much greater by making use of whole-cell homogenates. These can be prepared from frozen stored as well as from freshly collected cells.
4. Accumulation of PMN in the peritoneal inflammatory exudate is greatest at this time (approx 1–2 × 10^9 PMN/exudate) and PMN still comprise >85% of the cells in the exudate. Animals should be injected two to three times at 1- to-2-wk intervals with approx 50–100 mL of glycogen/saline before the initial collection to prime the response. After initiating collection of exudates, animals are challenged at 2- to 3-week intervals and yield exudates that are highly reproducible in cellular yield and content.

5. Maximum solublization of BPI during acid extraction requires quantitative lysis of cells during homogenization. The extent of cellular break-up can be readily monitored by phase-contrast microscopy.
6. Dialysis is accompanied by progressive accumulation of (protein) precipitates as the pH of the dialyzed material is raised. Rabbit PMN extracts can be neutralized to pH 7–7.4 by dialysis vs 2 m*M* Tris-HCl at pH 7.4 with little or no loss of BPI. In contrast, human PMN acid extracts can not be dialyzed to pH > 4.0 without losing BPI by coprecipitation with other precipitating substances during dialysis to higher pH. Thus, human PMN extracts are routinely dialyzed against 10 m*M* acetate buffer at pH 4.0. Dialyzed human PMN extracts are stable for weeks at 4°C at pH 4.0, whereas rabbit extracts are more stable at pH 7.0, possibly reflecting the greater prominence of acid proteases in rabbit extracts and neutral proteases in human extracts.
7. Highest-affinity bacterial binding of native BPI is to Gram-negative bacteria containing LPS with short saccharide chains (i.e., R-LPS) ***(15,16)***. This includes K-12 strains of *E. coli*. Initial interaction of BPI with the Gram-negative bacterial surface is likely to anionic moieties in and adjacent to the highly conserved lipid A region of LPS ***(16,17)***. In the bacterial envelope, these sites are at the outer membrane interface and less accessible to BPI when LPS molecules contain long polysaccharide chains. In contrast, the presence of an external capsular polysaccharide layer does not impede BPI interactions with the bacterial outer membrane ***(18)***.
8. Optimal or near-optimal binding of BPI to *E. coli* occurs at 20–37°C over a broad range of pH (4.0–7.5) and salt concentrations (0–200 m*M* NaCl) and in the presence of high concentrations of other leukocyte proteins ***(10)***. As bacterial interactions of many other cationic antibacterial proteins of PMN are salt-sensitive and antagonized by physiological salt concentrations ***(2)***, the selectivity of BPI binding to bacteria when PMN extracts and BPI-sensitive bacteria are mixed is generally greater in incubation mixtures containing 0.9% (w/v) NaCl.
9. The selectivity of BPI binding to bacteria is also enhanced when incubation mixtures contain sufficient BPI to saturate bacterial surface binding sites ***(12)***. At saturation, up to 2×10^6 molecules of BPI are bound per bacterium ***(10)***. Preliminary experiments may be carried out with 10^8 bacteria in 0.2 mL to determine the extract/bacteria ratio that gives the greatest purity and yield of affinity-purified BPI. Mg^{2+}-eluates (20 µL) can be analyzed by analytical reversed-phase HPLC or by sodium dodecyl sulfate–polyacrylamide gel electrophoresis (SDS-PAGE) after precipitation of proteins with ice-cold 10% trichloroacetic acid. In the latter case, precipitates are collected by sedimentation at 10,000*g* for 10 min at 4°C, washed twice with ice-cold 5% trichloroacetic acid and diethyl ether and dried under nitrogen. The precipitated proteins are solubilized by resuspension in SDS-PAGE (Laemmli) buffer and heating for 5 min in a boiling-water bath and analyzed by SDS-PAGE.
10. The anionic sites to which BPI bind in and near the lipid A region normally complex divalent cations (Mg^{2+}, Ca^{2+}) with µ*M* affinity ***(19)***. Up to 90% of

surface-bound BPI can be displaced from these sites with supraphysiological concentrations of these divalent cations. Maximum displacement occurs with 0.2 *M* $MgCl_2$ or $CaCl_2$ in acidic buffer with no detectable release of bacterial protein and very little release of bacterial phospholipids or LPS. Substitution of divalent cations with high concentrations (1.5 *M*) of NaCl causes nearly as much release of bound BPI.

11. BPI is eluted as a discrete peak at ca. 65% acetonitrile *(10,12)*. Human BPI elutes at lower acetonitrile concentration than does rabbit BPI. Protein elution is most sensitively monitored by measuring optical density (OD) at 214 nm. Such measurements also provide sensitive and quantitative assessment of protein recoveries with similar sensitivity for BPI and several other physically unrelated and dissimilar proteins.
12. To recover fully bioactive BPI, eluted BPI must be promptly dialyzed to remove the acetonitrile. If not, aggregates of BPI will form refractory even to dispersal by treatment with SDS. The protein is most stable if dialyzed against a weakly acidic buffer such as acetate buffer.

References

1. Elsbach, P., Weiss, J., and Levy. O. (1999) Oxygen-independent antimicrobial systems of phagocytes, in *Inflammation: Basic Principles and Clinical Correlates*, 3rd ed. (Gallin, J. I., Snyderman, R. A., and Nathan, C. A., eds.), Raven Press, New York, pp. 801–817.
2. Levy, O. (1996) Antibiotic proteins of polymorphonuclear leukocytes. *Eur. J. Haematol.* **56,** 263–277.
3. Gazzano-Santoro, H., Parent, J. B., Grinna, L., Horwitz, A., Parsons, T., Theofan, G., Elsbach, P., Weiss, J., and Conlon, P. J. (1992) High-affinity binding of the bactericidal/permeability-increasing protein and a recombinant amino-terminal fragment to the lipid A region of lipopolysaccharide. *Infect. Immun.* **60,** 754–761.
4. Elsbach, P. and Weiss, J. (1993) The bactericidal/permeability-increasing protein (BPI), a potent element in host-defense against gram-negative bacteria and lipolysaccharide. *Immunobiology* **187,** 417–429.
5. Weiss, J., Elsbach, P., Shu, C., Castillo, J., Grinna, L., Horwitz, A., and Theofan, G. (1992) Human bactericidal/permeability-increasing protein and a recombinant NH_2-terminal fragment cause killing of serum-resistant gram-negative bacteria in whole blood and inhibit tumor necrosis factor release induced by the bacteria. *J. Clin. Invest.* **90,** 1122–1130.
6. Rietschel, E. T. and Brade, H. (1992) Bacterial endotoxins. *Sci. Amer.* **267,** 54–61.
7. Abrahamson, S., Wu, H., Williams, R., Der, K., Ottah, N., Little, R., Gazzano-Santoro, H., Theofan, G., Bauer, R., Leigh, S., et. al. (1997) Biochemical characterization of recombinant fusions of lipopolysaccharide binding protein and bactericidal/permeability-increasing protein. Implications in biological activity. *J. Biol. Chem.* **272,** 2149–2155.

8. Weiss, J., Elsbach, P., Olsson, I., and Odeberg, H. (1978) Purification and characterization of a potent bactericidal and membrane-active protein from the granules of human polymorphonuclear leukocytes. *J. Biol. Chem.* **253,** 2664–2672.
9. Elsbach, P., Weiss, J., Franson, R., Beckerdite-Quagliata, S., Schneider, A., and Harris, L. (1979) Separation and purification of a potent bactericidal/permeability-increasing protein and a closely related phospholipase A_2 from rabbit polymorphonuclear leukocytes. Observations on their relationship. *J. Biol. Chem.* **254,** 11,000–11,009.
10. Mannion, B. A., Kalatzis, E. S., Weiss, J., and Elsbach, P. (1989) Preferential binding of the neutrophil cytoplasmic granule-derived bactericidal/permeability-increasing protein to target bacteria. Implications and use as a means of purification. *J. Immunol.* **142,** 2807–2812.
11. Weiss, J. and Olsson, I. (1987) Cellular and subcellular localization of the bactericidal/permeability-increasing protein of neutrophils. *Blood* **69,** 652–659.
12. Ooi, C. E., Weiss, J., Levy, O., and Elsbach, P. (1990) Isolation of two isoforms of a novel 15-kDa protein from rabbit polymorphonuclear leukocytes that modulate the antibacterial actions of other leukocyte proteins. *J. Biol. Chem.* **265,** 15,956–15,962.
13. Weinrauch, Y., Foreman, A., Shu, C., Zarember, K., Levy, O., Elsbach, P., and Weiss, J. (1995) Extracellular accumulation of potently microbicidal bactericidal/permeability-increasing protein (BPI) and p15s in an evolving sterile rabbit peritoneal inflammatory exudate. *J. Clin. Invest.* **95,** 1916–1924.
14. Zarember, K., Elsbach, P., Kim, K., and Weiss, J. (1997) p15s (15-kD antimicrobial proteins) are stored in the secondary granules of rabbit granulocytes: Implications for antibacterial synergy with the bactericidal/permeability-increasing protein in inflammatory fluids. *Blood* **89,** 672–679.
15. Weiss, J., Beckerdite-Quagliata, S., and Elsbach, P. (1980) Resistance of gram-negative bacteria to purified leukocyte proteins: relation to binding and bacterial lipopolysaccharide structure. *J. Clin. Invest.* **65,** 619–628.
16. Capodici, C., Chen, S., Sidorczyk, Z., Elsbach, P., and Weiss, J. (1994) Effect of lipopolysaccharide (LPS) chain length on interactions of bactericidal/permeability-increasing protein and its bioactive 23-kilodalton NH2-terminal fragment with isolated LPS and intact *Proteus mirabilis* and *Escherichia coli*. *Infect. Immun.* **62,** 259–265.
17. Weiss, J., Victor, M., and Elsbach, P. (1983) Role of charge and hydrophobic interactions in the action of the bactericidal/permeability-increasing protein of neutrophils on gram-negative bacteria. *J. Clin. Invest.* **71,** 540–549.
18. Weiss, J., Victor, M., Cross, A., and Elsbach, P. (1982) Sensitivity of K1-encapsulated *Escherichia coli* to killing by the bactericidal/permeability-increasing protein of rabbit and human neutrophils. *Infect. Immun.* **38,** 1149–1153.
19. Elsbach, P., Weiss, J., and Kao, L. (1985) The role of intramembrane Ca^{2+} in the hydrolysis of the phospholipids of *Escherichia coli* by Ca^{2+}-dependent phospholipases. *J. Biol. Chem.* **260,** 1618–1622.

5

Purification of Lipopolysaccharide-Binding Protein

Didier Heumann

Introduction

Lipopolysaccharide (LPS)-binding protein (LBP) is a critical component of innate immunity, implicated in the initiation of host defences against Gram-negative bacteria. LBP alerts the host to the presence of minute amounts of LPS *(1)*. LPS released from Gram-negative bacteria is present as aggregates, because of the amphiphilic structure of the molecule. LPS aggregates are transformed to monomers by the action of LBP, which has been described as a lipid-transfer molecule catalyzing movement of phospholipids including LPS *(2–6)*. When LPS/LBP monomers are transferred to lipoproteins, LPS is inactivated; when LPS/LBP complexes are transferred to cells harboring CD14 at their surface, cells are activated. Thus, the relative contribution of these two pathways will determine the response of the host to LPS.

Whereas LBP may have a protective role in alarming the host to minute doses of LPS, upon exposure to larger quantities of LPS, the amplification of LPS effects mediated by LBP may be detrimental to the host. Indeed, LBP has been shown to play a toxic role in experimental endotoxemia. Blockade of LBP activity with polyclonal and monoclonal antibodies was found to protect mice from lethal endotoxemia *(7,8,15a)*. Similarly, the disruption of the LBP gene has been associated with resistance to LPS *(9)*. If the role of LBP appears straightforward in endotoxemia, its contribution in fighting or helping infection is still largely unknown, inasmuch as LBP has been shown to protect mice from *Salmonella* infections.

There is a need to have well-characterized reagents to investigate the role of LBP as a mediator of activation of host cells by LPS or bacteria. This chapter summarizes protocols currently used in our laboratory for the purification of LBP and for the evaluation of LBP activity.

From: *Methods in Molecular Medicine, Vol. 36: Septic Shock*
Edited by: T. J. Evans © Humana Press Inc., Totowa, NJ

2. Materials

2.1. Sources of LBP

1. Recombinant LBPs: PCR-amplified murine *LBP* cDNA was cloned between the *Eco*RI and *Bgl*II sites of the baculovirus transfer vector pVL1392, as described previously *(10)*. Recombinant plasmid DNA, purified using Qiagen-tip 100 columns, was introduced into the Baculo-Gold virus (Pharmingen, San Diego, CA) using the cotransfection method. The recombinant virus was used to infect insect cells (either SF9 or SF21) cultivated in serum-free Excell medium. Infected cells were grown at 27°C in culture flasks and supernatant containing recombinant murine LBP (rmLBP) was collected after 4–5 d. The LBP construct adapted for the baculovirus system may be obtained from our laboratory. LBP supernatants may be stored at +4°C or –80°C for months before purification.

   Recombinant human LBP (rhLBP) may be obtained in the same manner using the baculovirus system *(11)*.

   In the present study, chinese ovary hamster cells (CHO cells) cells transfected with a DNA encoding human *LBP* were used as a source of rhLBP. *LBP* was inserted into the pRc/RSV vector (Invitrogen, San Diego, CA) containing a neomycin-resistance gene for selection of stable transformants and a Rous sarcoma virus (RSV) promoter for transcription of inserted DNA *(12,13)*. Electroporation was used to transfer the pRc/RSV constructs containing LBP into CHO cells, as described *(14)*. CHO cells transfected with rhLBP used in the present study were a kind gift of P.S. Tobias (Scripps Clinics, San Diego, CA).
2. Plasma: Human plasma was collected from human volunteers (use of citrate or heparin as anticoagulants does not affect results obtained). Serum may be an alternative source of LBP. Heparinized murine plasma is obtained by cardiac puncture.

2.2. Assay of LBP

1. Phosphate-buffered saline (PBS).
2. Antibodies to LBP: rat monoclonal and rabbit polyclonal to LBP (author's laboratory); murine monoclonal to human LBP (a kind gift of M. Miethke and W. Buurman, Universiteit Maastricht, Maasstricht, The Netherlands); rabbit polyclonal to human LBP (author's laboratory).
3. Blocking solution: 1% dried milk in PBS.
4. Peroxidase conjugates to rabbit or rat IgG (Sigma, Poole, Dorset, UK).
5. Microtiter plates (96-well) (Dynatech, Billingshurst, Sussex, UK).
6. Fluoroscein-labeled (FITC)-LPS (Sigma, Poole, Dorset, UK).
7. RPMI™ cell-culture medium (Life Technologies, Paisley, UK).

2.3. Purification of LBP

1. S-Sepharose (Pharmacia, St. Albans, Herts, UK).
2. Hi-Trap columns (Pharmacia).
3. Sodium acetate (20 m*M*) (NaAc), 0.4 *M* NaCl at pH 4.0.

4. Sodium acetate (20 m*M*) (NaAc), 1.5 *M* NaCl at pH 4.0.
5. HEPES buffer (50 m*M*) at pH 8.0.
6. Glycine HCl buffer (100 m*M*) at pH 2.8.
7. Tris 1 *M*—no adjustment to pH required.
8. Kits for protein concentration determination (Bio-Rad, Richmond, CA).

3. Methods

3.1. Assay of Murine LBP by ELISA

To follow the various steps of LBP enrichment during the process of purification, two methods are used. Enzyme-linked immunosorbent assay (ELISA) allows the quantification of the molecule, and a flow cytometric assay based on analysis of LBP-mediated binding of LPS labeled with FITC to $CD14^+$ cells allows the assessment of functional activity of the molecule. In addition, other functional tests may be performed that investigate the LBP-dependent production of cytokines (tumor-necrosis factor [TNF], for example) by monocytes or monocytic cell lines stimulated by LPS.

ELISAs can be performed using various combinations of antibodies. Polyclonal IgG to murine LBP can be used for coating plates, and LBP can be detected using the same preparation of IgG labeled with biotin, followed by the addition of streptavidin. In our laboratory, we use a rat monoclonal antibody recognizing rmLBP as a capture antibody and a rabbit anti-LBP IgG preparation to detect bound LBP ***(15a)*** or biotine-labeled rat anti-LBP mAB followed by avidine-peroxidase (not shown).

1. Coat 96-well microtiter plates with 1 µg/well MAb to rmLBP in 100 µL PBS, and incubate overnight at room temperature. Include negative control plates coated with PBS alone. All subsequent steps are performed at room temperature.
2. Incubate the plates with 100 µL of blocking solution for 1 h.
3. Remove blocking solution and wash three times with PBS.
4. Incubate 100 µL of samples to be analyzed in a range of dilutions using PBS/0.1% milk for 1 h. Include a standard curve using a range of known concentrations of purified LBP from any of the sources described above.
5. Wash three times with PBS.
6. Incubate with 100 µL of rabbit IgG to LBP (10 µg/mL) for 1 h.
7. Wash three times with PBS.
8. Incubate for 1 h with peroxidase-conjugated anti-rabbit IgG (1 in 1000 dilution) for 1 h.
9. Wash three times with PBS.
10. Peroxidase activity is detected by addition of 100 µl of freshly prepared o-phenylenediamine (OPD) substrate solution. Allow the reaction to proceed until a clear standard curve is seen and expected positives clearly developed.
11. Stop the reaction by addition of 100 µL of sulphuric acid. Measure the absorbances at 490 nm using a plate reader, correcting the results for variations in

optical density of each well by simultaneous measurement of absorbance at 690 nm. **Figure 1A** shows a representative standard curve of purified rmLBP using this assay.

3.2 Assay of Human LBP by ELISA

The ELISA measuring human LBP is essentially similar to that for measuring murine LBP. In this case, plates are coated with a murine MAb to human LBP (a kind gift of M. Miethke and W. Buurman), and after addition of LBP-containing fractions, a polyclonal rabbit IgG to human LBP is used to detect LBP, followed by anti-rabbit IgG labeled with peroxidase. **Figure 1B** shows a representative standard curve of purified rhLBP using this assay.

3.3. Assay of LBP by Flow Cytometric Assay

LBP can be functionally assessed by its capacity to present LPS-FITC to any source of $CD14^+$ cells. $CD14^+$ cells can be human circulating monocytes, or murine or human macrophage cell lines. Convenient targets are CHO cells *(15)* transfected with CD14 (either murine or human), because of an increased level of CD14 expression in these cell lines.

1. Mix dilutions of fractions containing LBP with FITC-LPS *(16)* used at a final concentration of 2 μg/mL and 10^5 target cells for 1 h at 37°C in a final volume of 200 μL RPMI/1% albumin.
2. Wash cells twice with cold (4°C) PBS.
3. After washing, the fluorescence on CD14 positive cells is quantified using flow cytometry *(17)*.

3.4. Protocol for Purification of LBP

LBP has been purified from humans, rabbits, calves, and mice *(18–20)*. Starting material can be either normal plasma or serum, although acute-phase plasma or serum is an enriched source of LBP. Alternatively, starting materials are supernatants of recombinant molecules expressed in insect cells or CHO cells.

Initial attempts of purification were made using two consecutive steps of ion-exchange chromatography, namely starting with LBP enrichment through an anion-exchange Bio-Rex-70 resin (Bio-Rad), followed by fractionation of the Bio-Rex eluate using a Mono-Q Pharmacia HPLC column *(18)*.

Although Bio-Rex chromatography allows enrichment of LBP from plasma, the best and perhaps the most suitable ion-exchange resin for purification of LBP is S-Sepharose *(21)*. We describe below a general technique that we have adapted from the technology of G. Theofan and colleagues *(21)*, with slight modifications. This protocol may be used for any form of LBP (plasmatic or recombinant), from human or murine sources. Other sources of LBP have not been investigated.

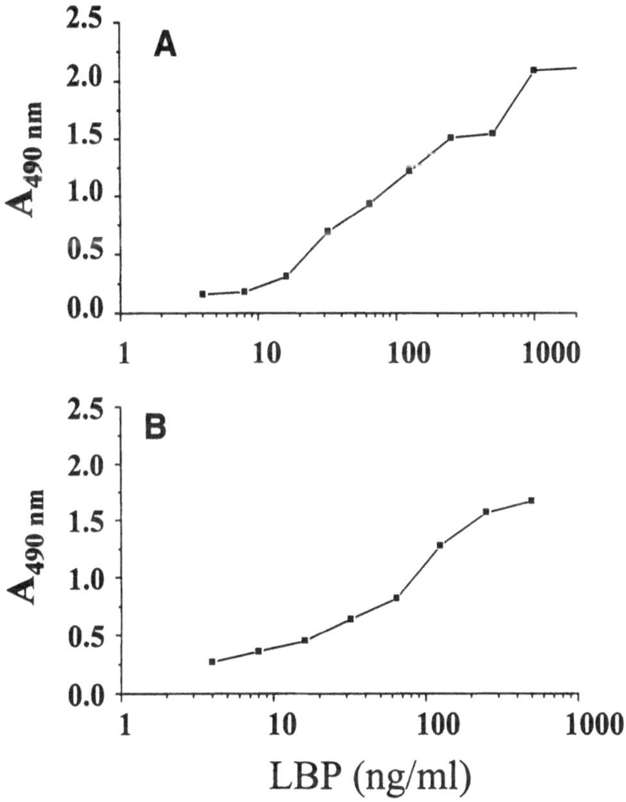

Fig. 1. ELISA for the determination of LBP concentrations. **(A)** Murine recombinant LBP, **(B)** recombinant human LBP.

3.4.1. Enrichment of LBP by Ion-Exchange Chromatography

1. Equilibrate S-Sepharose in 20 m*M* sodium acetate (NaAc), 0.4 *M* NaCl at pH 4.0 (*see* **Note 1**).
2. Dialyse LBP-containing material in 20 m*M* NaAc, 0.4 *M* NaCl at pH 4.0 overnight at 4°C.
3. Add 5% (v/v) of S-Sepharose beads equilibrated in buffer to the dialyzed material.
4. Mix by rotary inversion for 2 h at 4°C.
5. Pack the S-Sepharose into columns, and collect the flow-through to check the efficiency of the binding (*see* **Note 2**).
6. Wash extensively with the starting buffer until $A_{280} < 0.02$ (approx 10 column volumes).
7. LBP is batch eluted with the same buffer containing 1.5 *M* NaCl (*see* **Notes 3** and **4**).
8. Dialyze the S-Sepharose eluate against 50 m*M* HEPES buffer at pH 8.0 (*see* **Note 5**).

Table 1 shows the results of a typical purification from a supernatant of CHO cells transfected with rhLBP and from human plasma. LBP purification was assessed by ELISA, using purified rhLBP as standard. In this experiment, the LBP concentration in the CHO supernatant was 0.6 µg/mL (depending on the batches, LBP concentrations range from 0.2–1 µg/mL). Concentration of the CHO supernatant by S-Sepharose chromatography resulted in about a 15-fold increase in LBP concentration. LBP was recovered with an efficiency of 65%. Similar observations were made for human plasma as a source of LBP. The ion-exchange step reduced the protein concentration from 60 mg/mL in the starting material to 2 mg/mL, with a 69% recovery of LBP.

This technique was also adapted for the purification of murine LBP (**Table 2**). Starting material was a supernatant of insect cells infected with the baculovirus secreting rmLBP in the medium. In the example depicted in **Table 2**, 160 mL of SF9 supernatant was purified on S-Sepharose. This procedure reduced the volume to 9 mL. Half of the LBP content in the supernatant was present in the S-Sepharose eluate.

This procedure has also been used for the purification of LBP from plasma (data not shown).

3.4.2. Purification of LBP by Immunoaffinity Chromatography

Upon enrichment with S-Sepharose chromatography, LBP is not pure. As shown for example in **Table 2**, the S-Sepharose eluate still contained proteins other than LBP, as total proteins represented 720 µg (9 mL × 80 µg/mL), of which only 72 µg (9 mL × 8 µg/mL) was LBP. To further purify LBP, the most convenient procedure is the use of affinity chromatography using anti-LBP antibodies (*see* **Note 6**). Monoclonal antibodies to human or murine LBP are linked to Hi-Trap columns using basic protocols from Pharmacia.

1. Dialyse eluates of S-Sepharose beads against PBS.
2. Adsorb eluates onto an anti-LBP affinity column (equilibrated in PBS), at a flow rate of 1 mL/min, then extensively wash with PBS until no protein can be detected in the flow through ($A_{280} < 0.02$).
2. Elute LBP with 100 m*M* glycine HCl buffer at pH 2.8. Neutralize LBP with 1 *M* Tris. The flow-through is kept to check the efficiency of the procedure.
3. After elution, dialyze purified LBP against 50 m*M* HEPES at pH 8.0 (*see* **Note 7**)
4. Determine the amount of LBP by determining the protein concentration (BCA protein assay, Pierce, Rockford, IL, or Bio-Rad protein assay, BioRad Laboratories, München, Germany, using albumin as a standard.
5. Keep LBP at 4°C (*see* **Note 5**). This material serves as a standard for ELISA.

As an example, after the immunoaffinity procedure, LBP was pure at 90% (confirmed by SDS-PAGE). This represents 40% of the starting material, and

Table 1
Purification of Human LBP from Supernatants of CHO Cells Transfected with rhLBP and from Human Plasma

Material	Volume	Proteins	LBP/mL[a]	Total LBP	Efficiency of recovery[b]
CHO cells transfected with rhLBP:					
Supernatant	80 mL	64 µg/mL	0.6 µg/mL	48 µg	100%
Sepharose eluate	3.1 mL	50 µg/mL	10 µg/mL	31 µg	65%
Human plasma:					
Plasma	120 mL	60 mg/mL	3.5 µg/mL	420 µg	100%
Sepharose eluate	23 mL	2 mg/mL	12 µg/mL	288 µg	69%

[a]LBP concentration was determined by ELISA using immunoadsorbant-purified rhLBP, and checked for functional activity by the flow cytometric assay.

[b]Recovery of LBP as compared to the concentration of LBP measured in the starting material (supernatant or plasma).

Table 2
Purification of Recombinant Murine LBP from Supernatants of Insect Cells Infected with the Baculovirus Containing rmLBP (supernatant SF9)

Material	Volume	Proteins	LBP/mL[a]	Total LBP	Efficiency of recovery[b]
Supernatant	160 mL	50 µg/mL	1 µg/mL	160 µg	100%
Sepharose eluate	9 mL	80 µg/mL	8 µg/mL	72 µg	50 %
Anti-LBP imunoadsorbant eluate	1 mL	60 µg/mL	50 µg/mL	50 µg	40%

[a]LBP concentration was determined by ELISA using immunoadsorbant-purified rmLBP, and checked for functional activity by the flow cytometric assay.

[b]Recovery of LBP as compared to the concentration of LBP measured in the starting supernatant.

90% of the LBP from the S-Sepharose (**Table 2**; *see* **Note 8**). LBP activity in the flow-through of the immunoaffinity column was negligible, as measured by ELISA or by the flow-cytometric assay.

4. Notes

1. LBP has a tendency to stick nonspecifically to resins. In the initial procedure *(21)*, S-Sepharose was added directly to the medium containing LBP, in the absence of NaCl. In our hands, this procedure gave a strong adhesion of LBP to the resin. However, LBP could not be easily eluted from the resin. For these

Table 3
Concentrations of LBP after Ion-Exchange Chromatography Performed at Two Different Concentrations of Saline

	Buffer with 150 m*M* NaCl		Buffer with 400 m*M* NaCl	
	Proteins[a]	% LBP[b]	Proteins	% LBP[b]
Human plasma	12.6 mg/mL	0.04%	2 mg/mL	0.6%
Murine plasma	n.d.	n.d.	0.03%	0.4%
rhLBP(CHO supernatant)	200 μg/mL	4%	50 μg/mL	20%
rmLBP(SF9 supernatant)	320 μg/mL	2%	80 μg/mL	10%

[a]Concentration of proteins bound to the resin.
[b]% of LBP relative to total proteins bound to the resin (estimated by ELISA).
n.d.: Not determined.

reasons, we evaluated various concentrations of NaCl in the equilibration buffer. **Table 3** shows two examples of chromatography using 150 or 400 m*M* saline. The concentration of saline in the eluate modified the amount of proteins adsorbed to the column. In the presence of 150 m*M* NaCl, more proteins were bound to the resin than after using 400 m*M* NaCl. However, the relative amount of LBP in these proteins was higher using the 400 m*M* NaCl ionic strength. Furthermore, recovery of LBP was also better adding 400 m*M* NaCl in the equilibration buffer (not shown).

2. With the technique described here, LBP was virtually absent from the flow-through of the S-Sepharose column (as assessed by ELISA or by the flow-cytometric assay), indicating that losses of LBP were caused by nonspecific sticking to the resin and difficulty in eluting LBP from the column. Procedures devised to increase the yield of LBP beyond the limit of 50–70% have been unsuccessful in our laboratory.
3. Material eluted from the S-Sepharose beads is made up of many proteins in addition to LBP. For an example, in eluates originating from human plasma, LBP represents between 0.04 and 0.6% of the total proteins, depending on the ionic strength (**Table 3**). The concentration of LBP is more elevated in supernatants containing the recombinant molecules. Under these conditions, LBP in S-Sepharose eluate usually represents between 4 and 20% of proteins, depending on the origin of the culture medium and on the cell origin .
4. In the original procedure, Theofan and colleagues ***(21)*** suggested eluting LBP from S-Sepharose using a linear gradient of NaCl from 0.4–1 *M* NaCl in the sodium acetate buffer, followed by identification of fractions containing LBP by SDS-PAGE and pooling them. This procedure enriches LBP in purity, but additional procedures such as gel filtration or other ion-exchange columns are necessary to increase purity. Further purification steps may include gel filtration on Sephacryl S-100 equilibrated in 20 m*M* Na Citrate and 0.15 *M* NaCl at pH 5.0 or

Table 4
Ability of Murine LBP to Present LPS-FITC to CHO Cells Transfected with Murine CD14. The Effect of pH on the Stability of the Protein

	% of LBP activity relative to starting supernatant				
Dilutions	Starting supernatant	Eluate pH 4.0	Eluate pH 5.0	Eluate pH 7.0	Eluate pH 8.0
		Day 1			
1/20	100[a]	143	118	112	131
1/40	59	116	85	84	121
1/80	48	104	34	45	82
		Day 7			
1/20	100[a]	85	95	93	120
1/40	60	50	72	85	122
1/80	33	21	45	60	92
		Day 14			
1/20	100[a]	83	100	125	167
1/40	67	32	71	88	167
1/80	43	16	39	58	100

Supernatants and S-Sepharose eluates were diluted from 1/80 to 1/160 and added to CHO/CD14 cells together with LPS-FITC. Florescence on CHO/CD14 cells was measured.

[a]Fluorescence units were arbitrarily set to 100% for the 1/20 dilution of supernatant (standard). Numbers >100% indicate an increase of activity compared to the standard, numbers < 100% a decrease.

chromatography with mono-Q columns in 0.1 *M* phosphate buffer at pH 8.5, or a combination of both procedures, depending on the expected degree of purity.

5. There is a concern about the stability of purified LBP. In our hands, the stability of murine LBP and of human LBP differ. Human LBP (recombinant or from plasma origin) appears to be more stable over a large range of pHs.

   As shown on **Table 4**, there was a progressive loss of murine LBP activity over time depending on the pH of the buffer. The assay was normalized at 100% LBP activity for the starting supernatant, and the assay was performed the day after S-Sepharose chromatography, and 1 or 2 wk thereafter. On day 1, enrichment of LBP by chromatography on S-Sepharose resulted in an increased specific activity in the various eluates (as expected by the concentration step), independently of the pH of eluates. Roughly, a 5- to 10-fold higher dilution of eluate sustained a similar presentation of LPS-FITC as that obtained with the starting supernatant. However, the LBP activity was progressively lost over time in buffers made at pH 4.0, 5.0, and 7.0. The loss was the most important at pH 4.0. On day 14, only murine LBP kept at pH 8.0 retained its full activity. Unfortunately, from d 14 on, the activity of LBP was progressively lost, irrespective of the condition of the stock (–80°C or 4°C).

The functional activity of human LBP was similarly assessed. At pH ranging from 4.0–8.0, no significant loss was observed over time (data not shown).

The reasons for these differences in the stability of human and murine LBPs are not known. LBP has been found to be sensitive to proteases *(23)*. Other reasons could be change of conformation of the molecule or the aggregation of the protein. It was suggested that coating tubes with BSA would improve the stability of the protein. This procedure allows a better conservation of the molecule, but in our hands does not totally prevent the decrease in LBP activity over time. We found that freezing the protein (human and murine) at –80°C was associated with a loss of LBP activity over time. Importantly, polystyrene tubes should be avoided for LBP stocks. Polypropylene tubes (e.g., Eppendorf) could be used to limit nonspecific sticking of LBP to the walls. Before using LBP, it is mandatory to vortex the tubes extensively to allow LBP to get into solution.

6. When plasma is the source of material, LBP has to be enriched by ion-exchange chromatography before processing the material on affinity chromatography columns. This procedure eliminates approx 90% of plasma proteins, enriches quantitatively for LBP, and decreases nonspecific binding of plasma proteins to the immunoadsorbant. When supernatants of recombinant molecules are the source of material, the ion-exchange chromatography is not an obligatory step. However, independently of the starting material, the ion-exchange chromatography has the additional benefit of decreasing the volume of the starting material. For example, the ion-exchange chromatography step concentrates LBP from a volume of 80 mL (supernatant) to 3 mL (S-Sepharose eluate) (**Table 1**). The reduction of volume allows a simpler use of immunoadsorbants.
7. The steps of purification are followed by ELISA and functional assays that confirm that the purification steps do not lead to inactivation of the protein. The flow cytometric assay can also be used to quantify LBP concentrations *(22)*. In addition, the flow cytometric assay indicates that the various conditions of buffers used in the present procedure (high salt in the chromatographic step and acid buffer for elution of LBP from the immunoadsorbant) do not functionally alter LBP.
8. Using a two-step procedure consisting of ion-exchange chromatography and affinity chromatography using anti-LBP antibodies, it is possible to obtain purified LBP from various sources (supernatants or plasma). Because highly purified murine LP is not stable in the longterm, we use this form essentially for standardizing the amount of LBP in supernatants or enriched fractions eluted from the S-Sepharose column. It is indeed obvious that in most circumstances, semipurified LBP is sufficient, provided that the LBP concentration is known for these samples, and that other proteins do not prevent LBP activity. In the laboratory, we generally titrate supernatants of recombinant LBPs and use them for most of

our purposes. LBP activity in these supernatants is stable over months, does not depend on the pH of the buffers, and supernatants may be frozen at −80°C without loss of activity.

References

1. Schumann, R. R., Leong, S. R., Flaggs, G. W., Gray, P. W., Wright, S. D., Mathison, J. C., Tobias, P. S., and Ulevitch, R. J. (1990) Structure and function of lipopolysaccharide binding protein. *Science* **249,** 1429–1431.
2. Wurfel, M. M., Hailman, E., and Wright, S. D. (1995) Soluble CD14 acts as a shuttle in the neutralization of lipopolysaccharide (LPS) by LPS-binding protein and reconstituted high density lipoprotein. *J. Exp. Med.* **181,** 1743–1754.
3. Yu, B., Hailman, E., and Wright, S. D. (1997) Lipopolysaccharide binding protein and soluble CD14 catalyze exchange of phospholipids. *J. Clin. Invest.* **99,** 315–324.
4. Park, C. T. and Wright, S. D. (1996) Plasma lipopolysaccharide-binding protein is found associated with a particle containing apolipoprotein A-I, phospholipid, and factor H-related proteins. *J. Biol. Chem.* **271,** 18,054–18,060.
5. Wurfel, M. M. and Wright, S. D. (1997) Lipopolysaccharide-binding protein and soluble CD14 transfer lipopolysaccharide to phospholipid bilayers. Preferential interaction with particular classes of lipid. *J. Immunol.* **158,** 3925–3934.
6. Yu, B. and Wright, S. D. (1996) Catalytic properties of lipopolysaccharide (LPS) binding protein. *J. Biol. Chem.* **271,** 4100–4105.
7. Gallay, P., Heumann, D., Le Roy, D., Barras, C., and Glauser, M. P. (1993) Lipopolysaccharide-binding protein as a major plasma protein responsible for endotoxemic shock. *Proc. Natl. Acad. Sci. USA* **90,** 9935–9938.
8. Gallay, P., Heumann, D., Le Roy, D., Barras, C., and Glauser, M. P. (1994) Mode of action of anti-lipopolysaccharide binding protein (LBP) antibodies for prevention of endotoxemic shock in mice. *Proc. Natl. Acad. Sci. USA* **91,** 7922–7926.
9. Jack, R. S., Fan, X. L., Bernheiden, M., Rune, G., Ehlers, M., Weber, A., Kirsch, G., Mentel, R., Furll, B., Freudenberg, M., Schmitz, G., Stelter, F., and Schütt, C. (1997) Lipopolysaccharide-binding protein is required to combat a Gram-negative bacterial infection. *Nature* **389,** 742–745.
10. Lengacher, S., Jongeneel, C. V., Le Roy, D., Lee, J. D., Kravtchenko, V., Ulevitch, R. J., Glauser, M. P., and Heumann, D. (1996) Reactivity of murine and human recombinant LPS-binding protein (LBP) with LPS and Gram-negative bacteria. *J. Inflamm.* **47,** 165–172.
11. Goldblum, S. E., Brann, T. W., Ding, X., Pugin, J., and Tobias, P. S. (1994) Lipopolysaccharide (LPS)-binding protein and soluble CD14 function as accesory molecules for LPS-induced changes in endothelial barrier function, in vitro. *J. Clin. Invest.* **93,** 692–702.
12. Lee, J. D., Kato, K., Tobias, P. S., Kirkland, T. N., and Ulevitch, R. J. (1992) Transfection of CD14 into 70Z/3 cells dramatically enhances the sensitivity to complexes of lipopolysaccharide (LPS) and LPS binding protein. *J. Exp. Med.* **175,** 1697–1705.

13. Lee, J. D., Kravchenko, V., Kirkland, T. N., Han, J., Mackman, N., Moriarty, A., Leturcq, D., Tobias, P. S., and Ulevitch, R. J. (1993) Glycosyl-phosphatidyl-anchored or integral membrane forms of CD14 mediate identical cellular responses to endotoxin. *Proc. Natl. Acad. Sci. USA* **90,** 9930–9934.
14. Tobias, P. S., Soldau, K., Gegner, J. A., Mintz, D., and Ulevitch, R. J. (1995) Lipopolysaccharide binding protein-mediated complexation of lipopolysaccharide with soluble CD14. *J. Biol. Chem.* **270,** 10,482–10,488.
15. Delude, R. L., Fenton, M. J., Jr., Savedra, R., Perera, P. Y., Vogel, S. N., Thieringer, R., and Golenblock, D. T. (1994) CD14-mediated translocation of nuclear factor-κB induced by lipopolysaccharide does not require tyrosine kinase activity. *J. Biol. Chem.* **269,** 22,253–22,260.
15a. Le Roy, D., Di Padova, F., Tees, R., Lengacher, S., Landmann, R., Glauser, M. P., Calandra, T., and Heumann, D. (1999) Monoclonal antibodies to murine lipopolysaccharide (LPS)-binding protein (LBP) protect mice from lethal endotoxemia by blocking either the binding of LPS to LPB or the presentation of LPS/LBP complexes to CD14. *Journal of Immunology* **162,** 7454–7460.
16. Skelly, R. R., Munkenbeck, P., and Morrison, D. C. (1979) Stimulation of T-independent antibody responses by hapten-lipopolysaccharides without repeating polymeric structure. *Infect.Immun.* **23,** 287–293.
17. Heumann, D., Gallay, P., Barras, C., Zaech, P., Ulevitch, R. J., Tobias, P. S., Glauser, M. P., and Baumgartner, J. D. (1992) Control of LPS binding and LPS-induced TNF secretion in human peripheral blood monocytes. *J. Immunol.* **148,** 3505–3512.
18. Tobias, P. S., Soldau, K., and Ulevitch, R. J. (1986) Isolation of a lipopolysaccharide-binding acute phase reactant from rabbit serum. *J. Exp. Med.* **164,** 777–793.
19. Gallay, P., Carrel, S., Glauser, M. P., Barras, C., Ulevitch, R. J., Tobias, P. S., Baumgartner, J. D., and Heumann, D. (1993) Purification and characterization of murine LPS binding protein. *Infect. Immun.* **61,** 378–383.
20. Khemlani, L. S., Yang, Z., and Bochsler, P. N. (1994) Identification and characterization of a bovine lipopolysaccharide-binding protein. *J. Leuk. Biol.* **56,** 784–791.
21. Theofan, G., Horwitz, A. H., Williams, R. E., Liu, P. S., Chan, I., Birr, C., Carroll, S. F., Meszaros, K., Parent, J. B., Kasler, H., Aberle, S., Trown, P. W., and Gazzano-Santoro, H. (1994) An amino-terminal fragment of human lipopolysaccharide-binding protein retains lipid A binding but not CD14-stimulatory activity. *J. Immunol.* **152,** 3623–3629.
22. Heumann, D., Gallay, P., Le Roy, D., and Glauser, M. P. (1994) Radioimmunoassay versus flow cytometric assay to quantify LPS-binding protein (LBP) concentrations in human plasma. *J. Immunol. Methods* **171,** 169–176.
23. Dubin, W., Martin, T. R., Swoveland, P., Leturcq, D. J., Moriarty, A. M., Tobias, P. S., Bleicker, C. A., Goldblum, S. E., and Hasday, J. D. (1996) Asthma and endotoxin: lipopolysaccharide-binding protein and soluble CD14 in broncho-alveolar compartment. *Am. J. Physiol.* **270(5 Pt1),** L736–L744.

2

Other Bacterial Products

6

Purification of Streptococcal Pyrogenic Exotoxin A

Manuela Roggiani and Patrick M. Schlievert

1. Introduction

Group A streptococci secrete a variety of molecules, many of which are recognized as virulence factors important in the establishment of streptococcal infections. Among these extracellular products is streptococcal pyrogenic exotoxin A (SPE A, scarlet fever toxin A, erythrogenic toxin A) *(1)*. Other SPEs include toxin serotypes B and C *(1)*, streptococcal superantigen (SSA) *(2)*, and SPE F (mitogenic factor) *(3,4)*. Combinations of these toxins are believed to be important in streptococcal toxic shock syndrome. The latter two molecules will not be discussed in this chapter, but methods utilized to purify SPEs A–C also may be used to purify SSA and SPE F.

SPE A, B, and C are members of the pyrogenic toxin superantigen (PTSAg) family, a group of toxins of streptococcal and staphylococcal origin that share structural and biological properties. The PTSAgs are superantigens, and thus, induce a rapid and massive expansion of some subsets of the host T-cell population, accompanied by overproduction of a number of lymphokines and monokines, and followed in some instances by deletion of the stimulated T-cell subsets. Moreover, the PTSAgs are potent pyrogens, they enhance host susceptibility to endotoxin, they cause suppression of B-cell responses, and they are lethal to humans in 1–5 μg amounts. In addition to exhibiting these properties, SPE B is a cysteine protease and can cleave IL-1β precursor, vitronectin and fibronectin *(5)*.

The SPEs are true exotoxins, released from the bacterial cell upon cleavage of a signal peptide, and found in culture supernates. They are small proteins and are readily soluble in water, saline or mildly acidic aqueous solutions (**Table 1**). The two major isoforms of SPE A, of p*I* of 4.5 and 5.5 (**Table 1**) *(6)*,

From: *Methods in Molecular Medicine, Vol. 36: Septic Shock*
Edited by: T. J. Evans © Humana Press Inc., Totowa, NJ

Table 1
Biochemical Properties of the SPEs A–C

Protein	Molecular Weight	p*I*	Solubility	References
SPE A	25,787	4.5, 5.5	water, acetate-buffered saline, pH 4.5	*1,11,12*
SPE B	40,314 (inactive) 27,588 (active)	8.0, 8.3, 9.0	water acetate-buffered saline, pH 4.5	*1,7,13*
SPE C	24,354	6.7 or 7.0	water acetate-buffered saline, pH 4.5	*1,14,15*

can be reduced to one form by β-mercaptoethanol. Of the three forms of SPE B, only the form with p*I* 8.3 appears to be biologically active *(7)*. There is only one molecular form of SPE C.

The purification procedure performed in our laboratory for the SPE A–C consists of precipitation of the proteins from culture fluids by ethanol, resolubilization in pH 4.5 acetate-buffered saline or water, and separation of the proteins by successive flat-bed isoelectric focusing procedures (**Fig. 1**). SPE A, B, and C are generally purified from different strains of group A streptococci (**Table 2**), or as recombinant proteins from *Escherichia coli* (**Table 2**). In addition, SPE A is expressed and purified from *Staphylococcus aureus (1)* and *Bacillus subtilis (8)*, although in the latter case, a large portion of the secreted protein becomes irreversibly precipitated. On the contrary, SPE B and SPE C are not expressed in *S. aureus (1)*, and expression in *B. subtilis* has not been pursued. This report focuses primarily on purification of SPE A, but also lists the strains and methods used for purification of SPEs B and C.

2. Materials

1. Medium: The medium used for toxin production is a dialysate of trypsinized beef heart. For the medium preparation, commercially available beef hearts are ground with a standard meat grinder and heated to 70°C for 15–30 min in tap water (3 L of water/5 pounds of beef). The mixture is cooled to 40°C, and trypsin (16 g, Difco Laboratories, Detroit, MI; 1:250) is added. During trypsin treatment the pH is monitored and kept at 8.0 by addition of 5 or 10 *M* NaOH. Trypsinization usually requires 1–2 h. The digested beef is then poured into dialysis tubing with

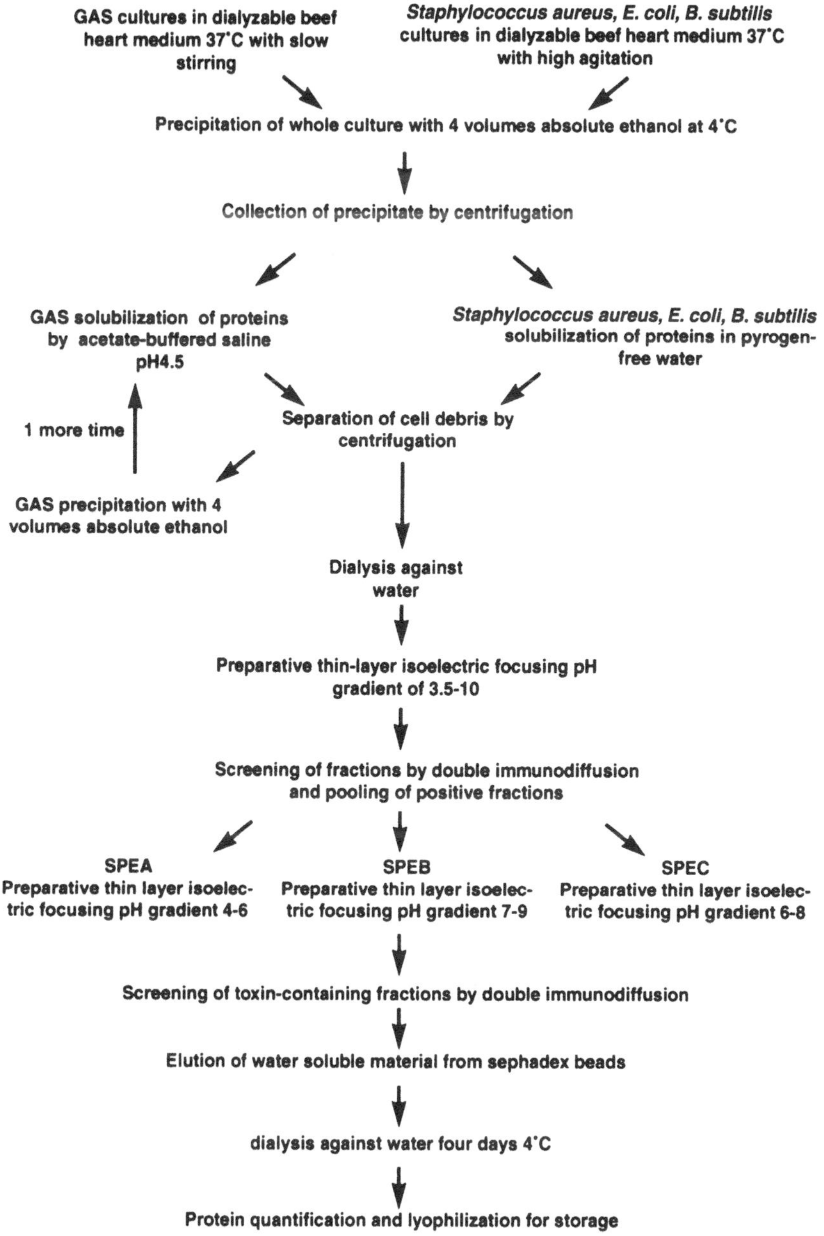

Fig. 1. Flow chart for purification of SPEs from group A streptococci (GAS), *S. aureus*, *E. coli*, or *B. subtilis*.

Table 2
Strains Commonly Used for Purification of SPEs A–C

Protein	Organism	Strain/Plasmid	Vector	Reference
SPE A	Group A streptococci (GAS)	594		*6*
		$T25_3$cured(T12)		*16*
	E. coli	RR1/pUMN103	pBR328	*16*
	B. subtilis	IS75/pUMN130	fusion of pBR328+pBD64	*8*
	S. aureus	RN4220/pMIN165	fusion of pBR328+pE194	*1*
SPE B	GAS	86-858		*13*
	E. coli	RR1/pUMN701	pBR328	*13*
SPE C	GAS	T18P		*1*
	E. coli	RR1/pUMN521	pUC19	*14,15*

a molecular weight cutoff of 12,000–14,000, and dialyzed for 4 d against distilled water (20 L). The dialysate is then aliquoted and autoclaved. The medium can be stored at room temperature for months.

2. Glucose phosphate buffer: 0.33 *M* glucose, 0.5 *M* $NaHCO_3$, 0.12 *M* Na_2HPO_4, 0.68 *M* NaCl, and 0.027 *M* L-glutamine. The buffer is filter sterilized (pore size 0.2 µm, Corning Costar, Corning, NY), and stored at 4°C. This solution is added to beef heart medium at 5% (v/v) for culturing streptococci, and at 1% (v/v) for culturing *S. aureus*, *E. coli*, and *B. subtilis*, in all cases just prior to use.
3. Antibiotics: erythromycin or chloramphenicol is added to the medium at 5 µg/mL, ampicillin at 100 µg/mL, for growth of *S. aureus*, *B. subtilis*, and *E. coli* clones, respectively.
4. Flat-bed isoelectric focusing preparative gels. Sephadex G75 (Sigma, St. Louis, MO) is used for solid support. Prior to use, the beads are swollen in distilled water, then washed with absolute ethanol to remove fine particulate material, and allowed to air dry. Ampholytes for the various pH gradients are purchased from Pharmacia Biotech (Uppsala, Sweden). Each preparative gel is prepared with 2.5 mL of ampholyte solution.
5. Acetate-buffered saline: 0.15 *M* NaCl in 5 m*M* sodium acetate at pH 4.5. The buffer is autoclaved, and stored at 4°C.
6. Polyclonal antibodies against each toxin separately are obtained from rabbits hyperimmunized with pure preparations of the specific toxin. Rabbits are hyperimmunized by subcutaneous administration in the nape of the neck of 50 µg toxin emulsified in incomplete Freund's adjuvant (Difco), on each of days 0, 14, 28, and 56. Animals are bled from the ear central artery one week after the last injection, and serum is collected.

7. Gels for double immunodiffusion are prepared with 0.75% (w/v) agarose (Sigma) in phosphate-buffered saline at pH 7.2 (5 m*M* Na_2HPO_4/NaH_2PO_4 buffer at pH 7.2, 150 m*M* NaCl). The melted agarose is poured onto glass slides (4.5 mL/slide), allowed to solidify, and stored at 4°C until used.

3. Method

An outline of the purification protocol for the SPEs followed in our laboratory is shown in **Fig. 1** (*see* **Note 1**).

1. Streptococci are grown in dialyzable beef heart medium completed with 5% (v/v) glucose phosphate-buffered solution, at 37°C, with light stirring (approx 25 rpm). Cultures are sequentially transferred 1/10 (v/v) into 100, 1000, and 10,000 mL prewarmed medium, and at each step grown to stationary phase. *S. aureus*, *E. coli*, or *B. subtilis* are grown in the same medium, except the medium is completed with 1% (v/v) glucose phosphate buffer, and the appropriate antibiotics for maintaining plasmids are added. Cultures are grown at 37°C, with vigorous agitation (100–200 rpm), for 24 h (well into stationary phase).
2. Culture fluid containing the native or the recombinant version of the desired toxin is slowly poured into 4 vol of absolute ethanol at 4°C. Precipitation of toxin is allowed to occur for at least 2 d at 4°C.
3. At the end of this time, the precipitate has settled to the bottom of the flasks. Most of the supernate is poured out, and the remainder is separated from the precipitated material by centrifugation (250*g*). Pellets are drained and allowed to air dry.
4. Pellets are resuspended in small volumes of acetate buffered saline for group A streptococci (usually 500 mL/10 L worth of original culture fluid), or 75 mL of distilled water for *S. aureus*, *E. coli*, or *B. subtilis* (usually 75 mL/5 L of original culture fluid).
5. The resolubilized portion is separated from the cell debris by centrifugation (10,000*g* for 30 min). Supernates are reprecipitated with ethanol and resolubilized in acetate-buffered saline for group A streptococci.
6. Supernates from *S. aureus*, *E. coli*, or *B Subtilis* in **step 5** are dialyzed overnight against a large volume (approx 40×) of distilled water, in dialysis tubing with a molecular weight cutoff 12,000–14,000. This step separates the toxin from most residual medium peptides and salts. For group A streptococci, solubilized material from the second ethanol precipitation is centrifuged at 10,000*g* and dialyzed 3 d against water (approx 40× each day).
7. The protein preparation from group A streptococci is lyophilized and dissolved in a small volume of water (150 mL/10 L of original culture fluid) for preparative isoelectric focusing. The toxin preparations from *S. aureus*, *E. coli*, or *B. subtilis* are air dried inside the dialysis tubing until they reach the volume of approx 75 mL.

8. Isoelectric focusing separation is performed on a preparative Sephadex G75 bed, first in pH gradient of 3.5–10. The proteins are allowed to focus for 18 h.
9. Fractions (1-cm wide) are harvested and suspended in a small volume (7–10 mL) of distilled water for each fraction. After the Sephadex G75 beads have settled to the bottom of the test tubes, the aqueous solutions from each fraction are tested in a double immunodiffusion assay (**step 10**, below), for detection of the toxin, by use of polyclonal antibodies raised against each toxin separately. The toxin-positive fractions are pooled.
10. Double immunodiffusion is performed by punching 4-mm diameter wells 4 mm from each other in a hexagonal pattern. Twenty microliters of toxin or antisera are added to the wells, and the slides are observed after 4 h at 37°C or overnight at 4°C for precipitin areas, indicative of toxin reactivity with antiserum.
11. The pooled fractions undergo a second preparative isoelectric focusing, in a narrower pH range. SPE A is focused in a pH gradient of 4.0–6.0, SPE B in a pH gradient of 7.0–9.0, and SPE C in a pH gradient of 6.0–8.0 (**Table 1**).
12. Fractions are again suspended in water and tested for reactivity with antisera in double immunodiffusion.
13. The toxin preparation from positive fractions is eluted from the Sephadex G75 matrix by passage through a disposable plastic syringe equipped with a glass wool plug. The eluted protein solution (15–20 mL) is dialyzed against distilled water in tubing with a molecular weight cutoff 6000–8000. Dialysis is performed for 4 d, with a change of water (1 L) every day, at 4°C. This extensive dialysis process is necessary to extract ampholytes from the protein preparation.
14. The purity of the protein preparation is assessed by SDS–polyacrylamide gel electrophoresis, followed by visualization by Coomassie blue R250 or silver staining of the gel slab (*see* **Note 2**).
15. The SPEs are best stored lyophilized. In this form they are stable at 4°C for over a year, and at −20°C for many years.
16. For some experimental purposes the SPEs must be separated from contaminants such as lipopolysaccharide (preparations from *E. coli*, *see* **Note 3**) or protein A (preparations from *S. aureus*). Further purification is achieved by reverse phase HPLC, in an acetonitrile gradient of 0–60% (SPE A). When the gradient is spread over a time period of 40 min, SPE A is released from the 15-cm C18 column (VY DAC, Hesperia, CA) at approx 22–24 min. The acetonitrile present is eliminated from the toxin preparation by lyophilization.

4. Notes

1. Throughout this procedure, as well as in the preparation of medium and reagents, pyrogen-free water and pyrogen-free glassware (baked at 180°C for 3 h) are used. It is our experience that Sephadex G75 and ampholytes used in these procedures also are pyrogen free.
2. Group A streptococci yield 2–2.5 mg of toxin/L of culture fluid. *S. aureus*, *E. coli*, or *B. subtilis* yield 1–2 mg of toxin/L of culture fluid.

3. SPE A has been shown to bind lipopolysaccharide very tightly *(9)*. In light of the endotoxin enhancement property of SPE A *(10)*, even trace amounts of lipopolysaccharide contaminating the protein preparation may significantly affect the biological activity of the toxin batch. Nothing is known, thus far, about interactions of SPE B and C with lipopolysaccharide, although the possibility for these interactions to occur exists. For this reason, in our laboratory the SPEs destined for biological studies are routinely purified from Gram-positive bacteria.

References

1. Bohach, G. A., Fast, D. J., Nelson, R. D., and Schlievert, P. M. (1990) Staphylococcal and streptococcal pyrogenic toxins involved in toxic shock syndrome and related illnesses. *Crit. Rev. Microbiol.* **17,** 251–272.
2. Mollick, J. A., Miller, G. G., Musser, J. M., Cook, R. G., Grossman, D., and Rich, R. R. (1993) A novel superantigen isolated from pathogenic strains of *Streptococcus pyogenes* with aminoterminal homology to staphylococcal enterotoxins B and C. *J. Clin. Invest.* **92,** 710–719.
3. Yutsudo, T., Murai, H., Gonzales, J., Takao, T., Shimonishi, Y., Takeda, Y., Igarashi, H., and Hinuma, Y. (1992) A new type of mitogenic factor produced by *Streptococcus pyogenes. FEBS Lett.* **308,** 30–34.
4. Norrby-Teglund, A., Newton, D., Kotb, M., Holm, S. E., and Norgren, M. (1994) Superantigenic preoperties of the group A streptococcal exotoxin SPE F (MF). *Infect. Immun.* **62,** 5227–5223.
5. Musser, J. M. (1997) Streptococcal superantigen, mitogenic factor, and pyrogenic exotoxin B expressed by *Streptococcus pyogenes*, in *Superantigens* (Leung, Y. M., Huber, B. T., and Schlievert, P. M., eds.), Marcel Dekker, New York, pp. 281-310.
6. Nauciel, C., Blass, J., Mangalo, R., and Raynaud, M. (1969) Evidence for two molecular forms of streptococcal erythrogenic toxin. Conversion to a single form by 2-mercaptoethanol. *Eur. J. Biochem.* **11,** 160–164.
7. Barsumian, E. L., Cunningham, C. M., Schlievert, P. M., and Watson, D. W. (1978) Heterogeneity of group A streptococcal pyrogenic exotoxin type B. *Infect. Immun.* **20,** 512–518.
8. Kreiswirth, B. N., Handley, J. P., Schlievert, P. M., and Novick, R. P. (1987) Cloning and expression of streptococcal pyrogenic exotoxin A and staphylococcal toxic shock syndrome toxin-1 in *Bacillus subtilis. Mol. Gen. Genet.* **208,** 84–87.
9. Leonard, B. A. B. and Schlievert, P. M. (1992) Immune cell lethality induced by streptococcal pyrogenic exotoxin A and endotoxin. *Infect. Immun.* **60,** 3747–3755.
10. Kim, Y. B. and Watson, D. W. (1970) A purified group A streptococcal pyrogenic exotoxin. Physicochemical and biological properties including the enhancement of susceptibility to endotoxin lethal shock. *J. Exp. Med.* **131,** 611–628.
11. Johnson, L. P., Tomai, M. A., and Schlievert, P. M. (1986) Molecular analysis of bacteriophage involvement in group A streptococcal pyrogenic exotoxin A production. *J. Bacteriol.* **166,** 623–627.

12. Weeks, C. R. and Ferretti, J. J. (1986) Nucleotide sequence of the type A streptococcal exotoxin (erythrogenic toxin) gene from *Streptococcus pyogenes* bacteriophage T12. *Infect. Immun.* **52,** 144–150.
13. Hauser, A. R. and Schlievert, P. M. (1990) Nucleotide sequence of the streptococcal pyrogenic exotoxin type B gene and toxin relationship to streptococcal proteinase precursor. *J. Bacteriol.* **172,** 4536–4542.
14. Goshorn, S. C., Bohach, G. A., and Schlievert, P. M. (1988) Cloning and characterization of the gene, *speC*, for pyrogenic exotoxin type C from *Streptococcus pyogenes*. *Mol. Gen. Genet.* **212,** 66–70.
15. Goshorn, S. C. and Schlievert, P. M. (1988) Nucleotide sequence of streptococcal pyrogenic exotoxin type C. *Infect. Immun.* **56,** 2518–2520.
16. Johnson, L. P. and Schlievert, P. M. (1984) Group A streptococcal bacteriophage T12 carries the structural gene for pyrogenic exotoxin type A. *Mol. Gen. Genet.* **194,** 52–56.

7

Assay for Superantigens

Shiranee Sriskandan

1. Introduction

This chapter describes a quantitative assay for the streptococcal superantigen, streptococcal pyrogenic exotoxin A (SPEA), which can be used in broth, tissue-culture media, and certain sera. The protocol can be adapted to allow measurement of any bacterial superantigen or protein toxin, using different oligonucleotides to amplify the coding sequences from bacterial DNA. Previously, measurement of bacterial toxins such as SPEA was limited by poor test sensitivity and test samples could only be analyzed if concentrated. Techniques included Ouchterlony immunodiffusion or enzyme-linked immunosorbent assay (ELISA), although the lower limit of detection was 100-fold greater than that achievable in this system. *(1–3)* Recent methods described have achieved improved sensitivities, some using sandwich ELISA techniques, others using competitive ELISA techniques. *(4–8)*.

The method for sandwich ELISA is straightforward, and it is possible to generate the reagents required in the laboratory within a few months. The structural gene for the chosen toxin (in this case, SPEA) is amplified from bacterial DNA using the polymerase chain reaction (PCR). Recombinant toxin can be synthesized using commercially available bacterial expression systems. A polyclonal anti-toxin antibody is then raised in rabbits, and affinity purified. A sandwich ELISA can then be constructed, using the anti-toxin antibody to coat ELISA plates, the recombinant toxin as a standard, and biotinylated anti-toxin antibody as a third layer.

A second technique for bioassay of superantigenic activity will also be described, which exploits the pronounced proliferation of human T cells following superantigen exposure. Cultured human peripheral blood mononuclear

From: *Methods in Molecular Medicine, Vol. 36: Septic Shock*
Edited by: T. J. Evans © Humana Press Inc., Totowa, NJ

cells are stimulated using the test sample and mitogenesis is measured by quantitation of uptake of tritiated thymidine. This can verify that superantigen measured using ELISA is actually in a bioactive form, and not fragmented or altered structurally. In addition, it provides a relatively nonspecific way of measuring the total superantigenic activity in a given sample; this can be useful in situations in which there are a mixture of toxins or superantigens that may be interacting, or in situations in which a putative unknown toxin is sought. This method is not specific for superantigens, as it will detect any mitogen.

2. Materials

2.1. Extraction of Streptococcal Bacterial DNA for PCR

1. Group A streptococcal National Collection of Type Cultures strain 8198 (M1T1 SPEA+).
2. Todd Hewitt Broth (Oxoid, Basingstoke, UK).
3. Mutanolysin (Sigma, Poole, UK).
4. Solution 1: 50 m*M* glucose, 25 m*M* Tris-HCl (pH 8.0), 10 m*M* ethylenediamine tetraacetic acid (EDTA) (pH 8.0).
5. Solution 2: 0.2 *M* NaOH, 1% SDS; *must* be fresh.
6. PCI:phenol: chloroform: isoamyl alcohol (25:24:1).
7. CI:chloroform: isoamyl alcohol (24:1).
8. Ethanol.
9. Tris-EDTA (TE) buffer:10 m*M* Tris-HCl at pH 8.0, 1 m*M* EDTA.

2.2. Amplification of Structural Toxin Gene, spea

1. Taq polymerase and buffer with 1.5 m*M* $MgCl_2$ (Promega, Madison, WI)
2. dNTPs (500 μ*M*).
3. Paired oligonucleotides (0.2 μg/μL) from 5′ and 3′ ends of toxin structural gene with flanking restriction sites appropriate for vector, e.g., *Bam*H1-ended *spea*:

   5′-GGCGGATCCGCAACAAGACCCCGATCCA
   3′-GCGGATCCGCAGTAGGTAAGGTTGCCAA

2.3. Cloning and Expression of Recombinant SPEA (rSPEA)

1. Restriction enzymes: *Bam*H1, *Afl*1 (Gibco, Paisley, UK).
2. Expression vector, e.g., pET19b (Novogen, Madison, WI).
3. Calf intestinal alkaline phosphatase (GIBCO Paisley, UK).
4. *Escherichia coli* competent cells.
5. *E. coli* λDE3 lysogens of strain BL21.
6. Ampicillin.
7. Luria Bertani (LB) broth: 1% bacto-tryptone, 0.5% yeast extract, 1% NaCl at pH 7.5.
8. SOC broth: 2% bacto-tryptone, 0.5% yeast extract, 0.05% NaCl, 2.5 m*M* KCl at pH 7.0 (using NaOH). 10 m*M* $MgCl_2$, 20 m*M* glucose.

2.4. Polyclonal Antibody

1. Two half-lop rabbits.
2. Syringes (1-mL) and 23-gage needles.
3. Sterile 0.9% saline.
4. Purified native SPEA (Toxin Technology, Sarasota, FL) and rSPEA.
5. Freunds adjuvant (complete and incomplete, Sigma).
6. Coated immunosorbent plates (96-well) (Nunc, Life Technologies, Paisley, UK).
7. Horseradish peroxidase (HRP)-conjugated anti-rabbit immunoglobulin (e.g., donkey anti-rabbit IgG, Jackson Immunoresearch, West Grove, PEN, USA).
8. ELISA substrate for 20 mL: 8-mg o-phenylenediamine, 10 mL DDW, 4.9 mL 0.1 *M* citric acid, 5.1 mL 0.2 *M* $Na_2HPO_4.2H_2O$, 50 µL H_2O_2, added last. (Make fresh, protect from light).
9. Stop solution: 1 *M* H_2SO_4.
10. Plate reader with filter set at 492 nm.

2.5. Affinity Purification of Antibody

1. Cyanogen bromide-activated sepharose 4B beads (Pharmacia, St. Albans, UK).
2. 0.01 M HCl.
3. Coupling buffer: 0.1 *M* $NaHCO_3$ pH 8.3, 0.5 *M* NaCl.
4. Recombinant SPEA.
5. Column with sintered-glass filter: a 10-mL column for a 3-mL bed volume is ideal.
6. Blocking buffer: glycine 0.2 *M* at pH 8.0.
7. Acetate buffer: 0.5 *M* Na acetate at pH 4.3.
8. Storage buffer: PBS; 0.1 *M* sodium azide.
9. Washing buffer: PBS.
10. Anti-SPEA antiserum.
11. Stripping buffer: 0.1*M* glycine.HCl at pH 2.5.
12. Collection buffer: 1 *M* Tris-HCl at pH 8.0.
13. Sephadex desalting column.
14. Sterile saline.

2.6. Biotinylation of Antibody

1. 0.1 *M* Na borate buffer at pH 8.8.
2. Biotinamidocaproate N-Hydroxysuccinamide ester (Sigma).
3. Dimethyl sulfoxide (Sigma).
4. 1 *M* NH_4Cl.
5. PBS.

2.7. ELISA

1. Coated immunosorbent plates (96-well) (Nunc, Life Technologies, Paisley, UK).
2. Affinity-purified anti-SPEA (stored at –20°C until use).

3. Coating buffer: 0.05 *M* Na carbonate at pH 9.6 (stored at 4°C or with azide).
4. Wash buffer: PBS, 0.1% Tween 20.
5. Blocking solution: 1% bovine serum albumin (BSA) in PBS (make up fresh).
6. rSPEA/samples (stored at –20°C until use).
7. Diluent for standard curve (e.g., 20% TH broth in PBS or 20% serum in PBS)
8. Biotinylated anti-SPEA antibody (stored at –20°C until use).
9. HRP-conjugated anti-rabbit immunoglobulin (e.g., donkey anti-rabbit IgG, Jackson Immunoresearch).
10. ELISA substrate: for 20 mL, 8 mg o-phenylenediamine, 10 mL DDW, 4.9 ml 0.1 *M* citric acid, 5.1 mL 0.2 *M* $Na_2HPO_4.2H_2O$, 50 µL H_2O_2: Add last. (Make up fresh, protect from light).
11. Stop solution: 1 *M* H_2SO_4.
12. Plate reader with filter set at 492 nm.
13. Graph paper/computer software to analyze data.

2.8. Bioassay for Superantigenic Activity

1. Freshly drawn human blood. (20 mL)
2. Preservative-free heparin or other anticoagulant.
3. Ficoll-Paque (Pharmacia).
4. Sterile 0.9% saline.
5. Hanks balanced salt solution (Gibco).
6. Roswell Park Memorial Institute 1640 medium reconstituted with 1× glutamine, penicillin/streptomycin solutions (Gibco).
7. Fetal calf serum
8. Positive controls: rSPEA (or TSST-1, SEA, SEB, phytohemagglutinin: all from Sigma).
9. Tritiated thymidine 1 µCi/µL: total of 100 µCi needed for one plate of cells.
10. Scintillation fluid: Betascint
11. Equipment: 96-well tissue culture plates, cell harvester, filter mats/sealable bags for cell harvesting, Betacounter.

3. Methods

3.1. Extraction of Bacterial DNA from Streptococcus Pyogenes for PCR

This is a quick method that is guaranteed to yield enough crude genomic DNA from Gram-positive bacteria for PCR amplification of target sequences. It is essentially an alkaline-lysis method.

1. Grow up a 10 mL culture of *S. pyogenes* in Todd Hewitt broth overnight at 37°C, without agitation.
2. Pellet cells by centrifugation at 2000*g* for 10 min. Pre-incubate with mutanolysin at 37°C for 30 min.
3. Pour off all medium and add 200 µL of solution 1.

4. Vortex vigorously for 2 min; transfer to a 1.5-mL polypropylene tube on ice.
5. Add 400 µL of solution 2.
6. Mix by inversion and incubate on ice till contents are viscous
7. Add 600 µL of PCI (caution) and vortex briefly
8. Centrifuge at 12,000*g* for 5 min.
9. Remove upper aqueous phase to a fresh polypropylene tube; discard other tube.
10. If aqueous phase still appears cloudy, repeat **steps 7–9**. Then repeat **steps 7–9** with CI.
11. Add 10% (v/v) 3 *M* sodium acetate at pH 5.2, and precipitate DNA/RNA with 2 vol of ice-cold ethanol. Leave at –20°C for 15 min–overnight.
12. Centrifuge for 15 min at 12,000*g* and aspirate pellet to dryness.
13. Allow pellet to dry in air, then resuspend nucleic acids in 50 µL TE buffer.

3.2. Amplification of Structural Toxin Gene, spea

1. The PCR reaction should comprise: distilled water 69.5 µL, dNTPs 5 µL, 5′ primer 5 µL, 3′ primer 5 µL, 10× Taq polymerase buffer 10 µL (1.5 m*M* $MgCl_2$), Taq polymerase 0.5 µL, target DNA 5 µL from toxin-producing strain, with 100 µL of mineral oil overlay.
2. Denature at 94°C for 30 s, annealing at 58°C for 45 s, and elongation at 72°C for 2 min over 29 cycles (initial denaturing step of 94°C for 2 min, final elongation at 72°C for 10 min).
3. Purify the product using standard procedures, digest an aliquot with restriction enzyme to create sticky ends, compatible with chosen expression vector, then purify again (*Bam*H1 was used for *spea*).

3.3. Cloning and Expression of rSPEA

1. Ligate sticky-ended toxin gene insert to cut expression vector using standard techniques.
2. Transform *E. coli* with ligation reaction and select for transformants using appropriate antibiotics. At this stage it is necessary to check that the insert is orientated correctly for expression, unless two different restriction sites have been used for ligation (*see* **Note 1**).
3. Sequence the insert to check for errors in amplification.
4. Induce expression of toxin by transferring the construct to an appropriate host bacterial strain; this will usually involve addition of isopropyl-B-thio-galactopyranoside (IPTG).
5. Purify expressed protein as recommended for the expression system used and analyse by sodium dodecyl sulfate-polyacrylamide gel electrophoresis (SDS-PAGE) (*see* **Note 2**).

3.4. Polyclonal Antibody to SPEA

1. Mix purified native SPEA (100 µg in 0.25 mL) with 0.25 mL of Freund's complete adjuvant and administer it to a half-lop rabbit.

2. Perform five subsequent im immunizations with 10 µg of recombinant SPEA mixed with Freund's incomplete adjuvant at two weekly intervals.
3. Assess immunity to SPEA by modified ELISA. Coat wells of a 96-well microtiter plate overnight at 4°C with 50 µL of rSPEA, diluted to 50 µg/mL in PBS (1 µg/well).
4. Wash wells three times with wash buffer; block wells with blocking buffer for 2 h, then wash again.
5. Make serial 10-fold dilutions of pre- and postimmune rabbit serum in PBS and add 100 µL of each dilution to microtiter wells, in duplicate. Incubate for 1 h at room temperature, then wash three times.
6. Dilute donkey anti-rabbit IgG antibody conjugated to HRP 1:2000 and add 100 µL to wells. Incubate for 1 h at room temperature, then wash three times.
7. Add 100 µL/well peroxidase substrate, which leads to an oxidative yellow color change. Stop reaction with 100 µL/well 1 *M* H_2SO_4 and measure absorbance at 492 nm. Specific immunity to toxin should be evident at titres as low as $1:10^{10}$, compared with pre-immune serum.

3.5. Affinity Purification of Antibody

1. Reswell CNBr-activated beads by soaking in 0.01 *M* HCl for 20 min in 20-mL tube.
2. Pour off HCl; pour in coupling buffer to neutralize beads (pH 8.0).
3. Pour off coupling buffer, then resuspend beads in a 1–5 mg/mL solution of rSPEA made up in fresh coupling buffer.
4. Incubate mixture on rotary mixer overnight at 4°C.
5. Pour matrix into column and drain fluid.
6. Wash the column through with three vol of coupling buffer.
7. Resuspend the matrix in blocking buffer and incubate on rotary mixer for 2 h at room temperature.
8. Wash the column with 3 vol of acetate buffer, then 3 vol of coupling buffer and, finally, 1 vol of PBS. (The column can be stored at this stage with a 'head' of PBS/azide, at 4°C).
9. For affinity purification, wash column thoroughly with 10 column volumes of PBS.
10. At 4°C, add 3 vol of anti-serum to the column (e.g., 10 mL for a small 3-mL bed volume) and allow to drain by gravity. Pass eluate over the same column three times to ensure maximal binding.
11. Wash the column with PBS (1 mL/min). Monitor the optical density (OD) of the eluate at 280 nm: the OD should return to the baseline level, once all unbound proteins have been washed away.
12. Strip the column of bound antibody by passing glycine strip buffer through at 1 mL/min. Collect 1-mL fractions of the eluate into separate tubes, each containing 50 µL 1 *M* Tris-HCl at pH 8.0.
13. Measure the OD of each fraction at 280 nm (diluted, if necessary) and pool the fractions with the greatest OD.

14. Desalt the antibody using a Sephadex column, into saline or PBS.
15. The protein content of the affinity-purified antibody can then be determined using a standard protein assay.

3.6. Biotinylation of Antibody

1. Dialyze approx 1 mL of affinity-purified anti-SPEA into sodium borate buffer. The concentration of antibody should ideally be 1–2 mg/mL.
2. Make up biotinamidocaproate N-hydroxysucinamide ester to a concentration of 10 mg/mL in dimethyl sulfoxide. Add to antibody solution in a ratio of approx 100 µg ester per 1 mg antibody and incubate for 4 h at room temperature.
3. Add 10 µL of 1 *M* ammonium chloride per 100 µg of ester and incubate 10 min
4. Dialyze the biotinylated antibody into PBS thoroughly.

3.7. ELISA Protocol

The timescale for this procedure is 2 d, provided that plates are pre-coated. This allows overnight incubation of the samples.

1. Coat a 96-well plate with 100 µL/well anti-rSPEA (diluted 1:1000 in 0.05 *M* carbonate buffer) and incubate overnight at 4°C. (Note: Coated plates can be stored at 4°C for 2 wk).
2. Wash 3× with PBS/Tween.
3. Block plates with 250 µL/well blocking buffer for 2 h at RT (or for up to 5 d at 4°C).
4. Wash 3× with PBS/Tween.
5. Dilute test samples in PBS (e.g., 1 in 5; 20 µL sample + 80 µL PBS) and plate out 100 µL/well in duplicate or triplicate, in rows C–H of the 96-well plate.
6. Prepare a diluent for the standard curve. This should reflect the constituents of the test wells, i.e., 20% broth/medium/normal serum: 80% PBS. For one plate, prepare at least 3 mL of diluent.
7. Prepare standard concentrations of rSPEA using the diluent. Doubling dilutions should be made from around 3 µg/mL to 3 ng/mL.
8. Standard concentrations of rSPEA should then be plated out in duplicate, in volumes of 100 µL/well, and incubated at 4°C overnight.
9. Wash 3× with PBS/Tween.
10. Add 100 µL/well biotinylated anti-rSPEA (1:1000 dilution). Incubate for 1 h at RT.
11. Wash 3× with PBS/Tween.
12. Develop ELISA using 100 µL/well streptavidin-HRP conjugate (1:2000 dilution in PBS) for 1 h at room temperature, wash 3× with PBS/Tween, then add 100 µL substrate to each well.
13. Stop the reaction once a standard curve is clear with 100 µL/well 1 *M* H_2SO_4 and measure the absorbance at 492 nm.
14. The relation between OD_{492} nm and rSPEA concentration should be plotted (**Fig. 1**; *see* **Note 3**).

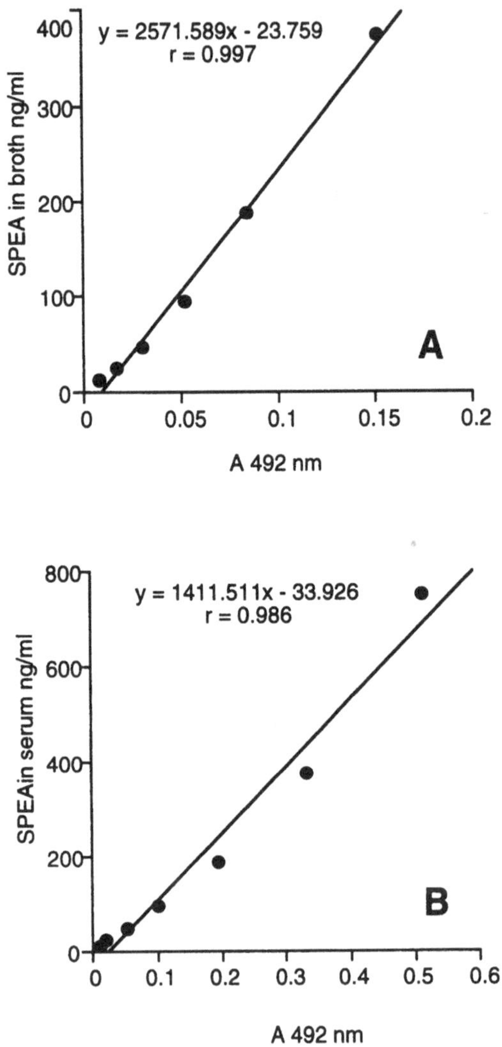

Fig. 1. Typical ELISA standard curve plots for **(A)** SPEA concentrations broth and **(B)** serum (A 492 nm: absorbance at 492 nm).

15. Values for SPEA concentrations in test samples can be mathematically derived from the A492 values, using a standard equation generated by appropriate software and corrected for dilution (*see* **Note 4**).

3.8. Bioassay for Superantigens

The time scale for bioassay using tritiated thymidine incorporation is 3.5 d. Cytokine production by peripheral blood mononuclear cells (PBMCs) after 72-h incubation has been used as an alternative parameter of superantigen bioactivity. Our laboratory has found that production of lymphotoxin-α (TNF-β) is a reliable indicator of T-cell activation, whereas others have measured IL-2 production. In general, however, T-cell proliferation is a more robust and reliable quantitative technique.

1. Draw 20 mL of fresh blood into a sterile 50-mL tube containing 100 µL heparin. The next steps should be performed in a tissue-culture flow hood.
2. Dilute blood twofold with sterile saline (40 mL total).
3. Put 15-mL Ficoll-Paque or similar into each of two 50-mL sterile tubes
4. Carefully layer 20 mL of diluted blood on top of the Ficoll-Paque, using a long tissue-culture pipet, holding the tube at a 45° angle, steadied at the base. The process may take several minutes; do not allow a stream of blood to penetrate the surface of the Ficoll-Paque. Repeat for other tube. The blood layer should be distinct from the clear Ficoll-Paque below. The protocol described below is a 'quick' version of that recommended by the manufacturers, but is adequate for the cell purity and numbers required for this method.
5. Carefully transfer the tubes to a centrifuge and spin at 800*g* for 25 min.
6. The lymphocyte layer should be visible as a gray line close to the interface of Ficoll and plasma. Aspirate the gray layer into a fresh 50-mL tube.
7. Resuspend the lymphocytes in a total of 50 mL of Hanks salt solution to wash.
8. Centrifuge at 400*g* for 10 min to pellet the cells.
9. Aspirate and discard most of the salt solution. Resuspend the lymphocyte pellet in 5–10 mL of tissue culture medium (made up with 10% fetal calf serum) by gently tapping the tube base and by pipetting up and down.
10. Count cells and add medium to give a final suspension of 10^6 cells/mL.
11. Plate 200 µL cell suspension per well (i.e., 2×10^5 cells/well). There should be enough cells for two plates.
12. Dilute positive control stimulants in tissue culture medium. Add to wells in volumes of 20 or 50 µL. Appropriate final concentrations to use for known superantigens are 1, 10, 100, and 1000 ng/mL. A formal dose response should be performed when testing these reagents for the first time with a new donor. Use PHA at 5 µg/mL. Each concentration should be tested in triplicate at least (the most convenient number of replicates will depend on the protocol used for beta counting).
13. Medium only is added to negative control wells. Negative control wells should be well separated from positive control wells.
14. Test samples (bacterial supernatants, serum, or purified protein—*see* **Note 5**) should be diluted and added to wells in the same volume of medium used for positive controls. Ensure that there are adequate numbers of replicates.
15. Plates are incubated for 72 h at 37°C in a standard tissue-culture incubator with 5% CO_2.
16. After plating cells (56–64 h), add 1 µCi/well of tritiated thymidine. This is most accurately done by diluting an aliquot of the stock solution of thymidine 1 in 20 using fresh medium; 20 µL/well is then added.

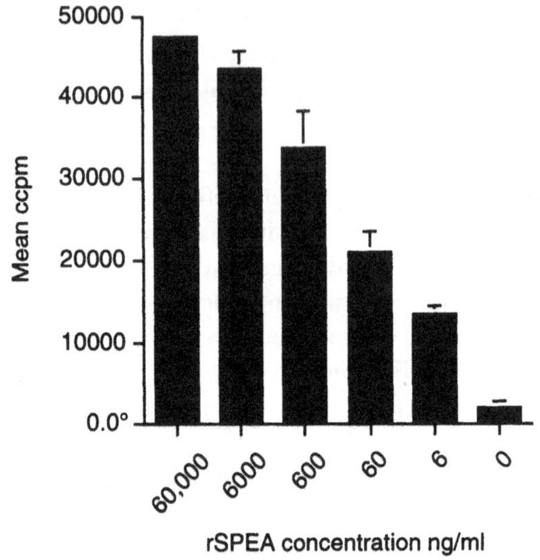

Fig. 2. Typical dose-response using rSPEA in bioassay. Stimulation of human mononuclear cell proliferation by rSPEA. Uptake of tritiated thymidine measured in (corrected) counts per minute (ccpm). Error bars show standard deviation of three experimental results.

17. Eight to 16 hours after addition of thymidine, plates are removed from the incubator.
18. Harvest cells onto filter mats; once dry and in sealable bag, add 10 mL of scintillation fluid, ensuring fluid is spread evenly across the mat. Beta counting can then be performed.
19. Counts from cells stimulated with test samples should be compared with counts from cells stimulated with known superantigens or mitogens (positive controls) and also compared with the negative controls (*see* **Note 6**). In general, positive controls will yield 20- to 100-fold increases in counts compared with negative controls, even when toxins are at nanogram concentrations (**Figs. 2** and **3**). Increases of lesser magnitude are unlikely to be caused by superantigenic activity.

4. Notes

1. For *spea* and pET19b, cut vector with *Xho*1 (immediately 5′ of the *Bam*H1 site), and then digest with *Afl*2, which cuts *spea* 180 bp from the 5′ primer start site. Electrophoresis of cut vector should reveal a band of approx 200 bp, confirming correct insert orientation.
2. In the case of pET 19b, the protein is tagged with a 5′ His-tag encoded by the pET vector which facilitates purification. For SPEA and pET 19b, a single 30-kDa protein is purified from solubilized inclusion bodies, corresponding to the predicted size of SPEA plus the His-tag. Up to 15 mg rSPEA has been extracted from 50-mL cultures, though the usual yield is 3–4 mg.

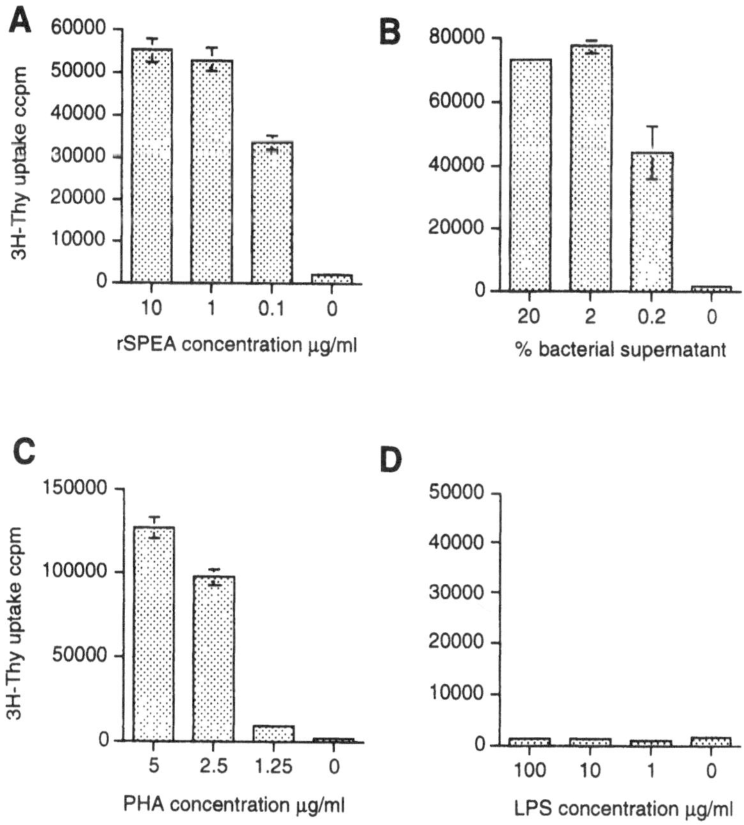

Fig. 3. Bioassay for superantigenic activity in broth. Proliferative response of human PBMCs to **(A)** rSPEA, **(B)** toxin-containing streptococcal supernatant, **(C)** PHA, and **(D)** LPS at concentrations shown. Bacterial supernatant provides potent stimulus to mitogenesis compared with pure toxin, presumably because it comprises several different superantigens acting in synergy. Mean of three experiments using single-donor PBMCs shown; error bars show standard errors.

3. The relation between rSPEA concentrations and the OD_{492} is normally linear in the concentration range 0–1 µg/mL rSPEA, but may yield a sigmoid curve if a wide concentration range of rSPEA is used. It is sometimes necessary to use semilog graph paper.
4. Sensitivity and specificity: the ELISA should be sensitive to 12 ng/mL rSPEA and has not been found to crossreact with non-SPEA-producing *S. pyogenes* supernatants, serum constituents from noninfected mice nor from animals with Gram-negative *E. coli* shock. Bacterial broths that contain high concentrations of immunoglobulin binding proteins (e.g., protein A from *Staphylococcus aureus*) can occasionally give false-positive results in the ELISA. This is evident when major color changes occur during the development stage of the ELISA, often

over and above the top of the standard curve. This may pose a theoretical problem with broths from *S. pyogenes*, which also produce IgG-binding proteins to a lesser degree, though, to date, we have not encountered this. If there is doubt as to the specificity of a positive result, the test sample should be subjected to SDS-PAGE and Western blotted, using the biotinylated anti-toxin antibody. This should reveal whether the ELISA is detecting a single protein of the desired molecular mass. If necessary, the test sample can be passed through an immunoglobulin column, to remove nonspecific Ig binding proteins.

Antibody crossreactivity with other staphylococcal toxins poses a theoretical problem, because of the considerable sequence similarity between streptococcal and staphylococcal superantigens. Murine IgM monoclonal antibodies against SPEA were found to crossreact with staphylococcal enterotoxin B, but this has not been tested using rabbit IgG polyclonal antibodies *(9)*. ELISA specificity can be further enhanced by use of anti-toxin antibody from an alternative source/species, either to coat plates, or as a third layer. If the antibody is not biotinylated and is used as a third layer, it would be necessary to replace the streptavidin-HRP step with a species-specific HRP-conjugated anti IgG antibody.

5. Bacterial supernatants should ideally have been prepared by growing bacteria in compatible tissue-culture medium (i.e., RPMI 1640 with glutamine and 10% serum, though without the antibiotics). Bacterial supernatants should not be added at greater than a 1 in 10 dilution, as they can be toxic. For superantigen activity in bacterial supernatnant, a 1 in 1000 dilution usually suffices. Serum from animals (infected with superantigen-producing bacteria or directly injected with superantigen) can also be tested for superantigenic activity, but controls should contain normal serum from the same species at the same concentration. Some sera will be toxic to the cultured cells, and some may contribute partly to mitogenesis. Concentrations of test serum to use will vary according to the experiment, but it is wise to test a range, e.g., 0.1–10%.
6. The mitogenesis test is wholly nonspecific and will detect any agent that can cause T-cell proliferation. To confirm that a substance is a superantigen, it is necessary to demonstrate that the T cells responding to stimulation express a limited repertoire of T-cell receptor (TCR) Vβ proteins. The response should occur without evidence of antigen processing, though normally requires class II molecules. For a known superantigen such as SPEA, the technique offers a simple bioassay to complement ELISA, immunodiffusion, or Western blotting data.

   Bacterial supernatants contain a number of superantigens and mitogens which act synergistically to cause T-cell proliferation. Thus, it is unrealistic to attempt to quantify the concentration of a single superantigen in broth using bioassay (*see* **Fig. 3**).

References

1. Houston, C. W. and Ferretti, J. J. (1981) Enzyme-linked immunosorbent assay for detection of type A streptococcal exotoxin: kinetics and regulation during growth of *Streptococcus pyogenes. Infect. Immun.* **33,** 862–869.

2. Kohler, W., Gerlach, D., and Knoll, H. (1987) Streptococcal outbreaks and erythrogenic toxin type A. *Zbl. Bakt. Hyg. A* **266,** 104–115.
3. Alouf, J. E., Knoll, H., and Kohler, W. (1991) The family of mitogenic, shock-inducing and superantigenic toxins from staphylococci and streptococci, in *Sourcebook of Bacterial Protein Toxins*. Academic Press, London, pp. 367–414.
4. Melish, M. E., Murata, S., Fukunga, C., Frogner, K., Hirata, S., and Wong, C. (1989) Endotoxin is not an essential mediator in toxic shock syndrome. *Rev. Infect. Dis.* **11 (suppl. 1),** S219–S228.
5. Mascini, E. M., Hazenberg, M. A. J., Verhage, L. A. E., Holm, S. E., Verhoef, J., van Dijk, H. (1996) A new procedure for the purification of streptococcal pyrogenic exotoxin A from *Streptococcus pyogenes* supernatant. *Clin. Diagnostic. Lab. Immunol.* **3,** 779–781.
6. Xu, S. and Collins, C. M. (1996) Temperature regulation of the streptococcal pyrogenic exotoxin A-encoding gene (*speA*). *Infect. Immun.* **64,** 5399–5402.
7. Sriskandan, S., Moyes, D., Buttery, L. K., Krausz, T., Evans, T. J., Polak, J., and Cohen, J. (1996) Streptococcal pyrogenic exotoxin A (SPEA) release, distribution and role in a murine model of fasciitis and multi-organ failure due to *Streptococcus pyogenes*. *J. Infect. Dis.* **173,** 1399–1407.
8. Hamad, A. R. A., Marrack, P., and Kappler, J. W. (1997) Transcytosis of staphylococcal superantigen toxins. *J. Exp. Med.* **185,** 1447–1454.
9. Bohach, G. A., Hovde, C. J., Handley, J. P., and Schlievert, P. M. (1988) Cross-neutralization of staphylococcal and streptococcal pyrogenic toxins by monoclonal and polyclonal antibodies. *Infect. Immun.* **56,** 400–404.

3

Cytokines

8

Bioassay for Tumor Necrosis Factors-α and β

Thomas J. Evans

1. Introduction

The realization that much of the toxicity of bacterial endotoxin resulted from production of a macrophage-derived intermediate *(1)* led to the isolation and cloning of tumor necrosis factor (TNF), also known as cachectin *(2)*. Since then, much evidence has accumulated to demonstrate that TNF is of prime importance in the pathogenesis of endotoxin-related tissue injury, and occupies a key proximal position in the cascade of mediators that are produced as a result of bacterial infection. These conclusions have been based in part on the accurate measurement of TNF in blood and other body fluids of humans and animals as a result of infection.

Prior to the purification of TNF, its measurement relied upon a bioassay *(3)*. This allowed detection of very low concentrations of TNF, as it has a high biological potency. However, in common with all bioassays, it is rather cumbersome to perform, and absolute results tended to vary from laboratory to laboratory. Potential interference from other bioactive substances might also make these assays not totally specific for TNF. Once the pure protein became available, a whole range of immunoassays was developed that could be performed easily, were highly specific, and were highly reproducible between different centres.

Why then does bioassay of TNF remain an important analytical tool? Firstly, immunoassay will measure total immunoreactive TNF, which may not necessarily be bioactive. The molecule may have become partially denatured and hence inactive, but the key epitope recognized in the immunoassay might remain intact. Immunoassay would then result in an overestimate of the amount of bioactive TNF present. Secondly, TNF does not circulate just as the free

From: *Methods in Molecular Medicine, Vol. 36: Septic Shock*
Edited by: T. J. Evans © Humana Press Inc., Totowa, NJ

active cytokine. Both the p55 and p75 TNF receptors are shed from cells in sepsis, and achieve high circulating levels *(4)*. They can bind TNF and sequester it in an intact but inactive form. Thus, certain immunoassays may measure relatively high levels of TNF but as it is bound to circulating receptors it is in fact biologically inactive. It is imperative, therefore, that if any meaningful biological conclusions are to be drawn about the levels of TNF in a study, their measurement must have accounted for the actual bioactive TNF that has been released. A good example of such a study was that performed relating TNF levels to outcome in patients suffering from meningococcal sepsis *(5)*.

Another form of tumor necrosis factor has been identified, termed lymphotoxin or TNF-β *(6)*. This shares approx 30% homology with TNF-α, and exerts many similar effects. This is because both cytokines bind to the same cellular receptors, the TNF p55 and p75 receptors. Most of the biological effects of TNF are mediated through the p55 receptor. There are important differences in the ability of TNFs from certain species to crossreact with receptors from another species. For example, murine TNF-α is able to bind and activate human p55 and p75 TNF receptors. However, human TNF can only bind the murine p55 receptor. TNF-β is a product of activated lymphocytes and in sepsis is generally produced in much lower amounts. However, it can be the predominant form of TNF released in certain circumstances, for example in the T-cell stimulation produced by superantigens that may account for the features of septic shock induced by these molecules *(7)*.

Bioassay of TNF relies on the ability of the molecule to produce cell death in certain target cell lines. A number of different human and murine lines have been used, which differ in their sensitivities towards TNF from different species. However, as long as the standard TNF employed is from the species under study, then the absolute values obtained will be biologically meaningful. Most of these assays employ agents to inhibit *de novo* RNA or protein synthesis, such as Actinomycin D. These act to prevent the synthesis of factors that protect against TNF cytotoxicity, such as manganese superoxide dismutase *(8)*. One potential source of error is the presence of factors that can also produce cell death in the assay, which are not related to TNF. Such factors might include other cytokines such as IL-1, interferon-γ, or lipopolysaccharide. Although there have been some reports of problems with these substances *(9)*, in practice published protocols have been careful to examine such potential interference from alternative cytotoxic agents.

The assay described here uses the cytotoxic effect of TNF on the murine fibrosarcoma line L929. It is a modification of a widely used protocol *(3,10)*. It can be adapted to measure TNF-α or TNF-β by the use of appropriate neutralizing antibodies, as described below. I describe our methods that we have used successfully for the assay of murine and human TNF.

2. Materials

2.1. TNF-α Assay

1. L929 cells: These are a murine fibrosarcoma cell line available from the European Collection of Animal Cell Cultures at Porton Down, Wiltshire, UK.
2. Tissue culture medium: RPMI 1640 medium supplemented with 10% fetal calf serum, penicillin 100 U/mL, streptomycin 100 µg/mL and glutamine 0.3 mg/mL.
3. Hank's buffered salt solution (HBSS) (Gibco-BRL, Paisley, UK).
4. TNF-α stock solution: Human and murine TNF are available from Peprotech, London, UK. Stocks should be aliquoted undiluted in 5-µL samples and stored at –80°C. Under these conditions, the cytokine is stable for at least 6 mo (*see* **Note 1**).
5. Actinomycin D: Dissolve chemical at 1 mg/mL in ethanol. Dilute this stock 20-fold into tissue-culture medium to give a final concentration of 50 µg/mL. Store 5-mL aliquots at –20°C. This is stable for at least 6 mo.
6. Trypsin/ethylenediamine tetraacetic acid (EDTA): 0.5 g/L trypsin, 0.2 g/L EDTA in modified Puck's saline A (Gibco BRL, Life Technologies, Paisley, UK).
7. Trypan blue: 0.4% solution in 0.9% saline.
8. Crystal violet: 0.5% solution in distilled water.
9. Formyl saline: 10% formaldehyde (v/v) in 0.9% saline. Make up fresh before use.
10. Flat bottomed microtiter plates (96-well) for tissue culture (Nunc, Gibco-BRL).
11. Neutralizing antibody to TNF-α, specific for species whose TNF is being tested. For murine and human TNF-α suitable neutralizing antibodies are available from Genzyme, West Malling, Kent, UK.

2.2 TNF-β Assay

1. Reagents as listed in **Subheading 2.1.** above.
2. Human recombinant TNF-β for standard curve (Genzyme). This is stable for 6 mo when frozen in aliquots at –80°C.

3. Methods

3.1. TNF-α Bioassay

3.1.1. Preparation of Cells

1. Prepare L929 cells: Perform all manipulations with growing cells in a tissue-culture flow hood. Grow cells until confluent in tissue-culture medium. Prepare enough cells to assay the numbers of samples you have. Enough cells can be isolated from a 75-cm^2 flask to coat two 96-well plates. Each plate can accommodate 60 samples, which must include 30 wells for the standard curve. This then allows 30 wells, enough to assay 10 samples in triplicate per plate (*see* **Note 2**).
2. Once cells are confluent, remove media from flask and wash with Hanks' buffered saline solution (HBSS).

3. Add 5 mL of Trypsin/EDTA per 75-cm^2 flask. Ensure all the cells are wetted with the solution, then remove nearly all the trypsin/EDTA.
4. Incubate the flask at 37°C until the cells begin to detach. This should take no longer than 3–4 min.
5. Rap the flask firmly with the palm of your hand and add 20 mL of tissue-culture fluid. Pipet the fluid up and down three times and then transfer to a sterile 25-mL universal container, with a V-shaped bottom.
6. Spin cells at 800*g* for 10 min at room temperature.
7. Discard supernatant, resuspend in 20-mL of tissue-culture medium and repeat centrifugation step.
8. Resuspend cells in 1 mL of tissue-culture medium.
9. Add 50 µL of cells into 400 µL of tissue-culture medium and add 50µL of trypan blue. Count cells using a hemocytometer and adjust cell suspension to 3×10^5 cells/mL.
10. Add 100 µL of the cell suspension (i.e., 3×10^4 cells) to each well of the flat-bottomed tissue-culture plate, leaving all the outside wells empty. This is because the outer well may suffer from evaporative loss and give rise to spurious cytotoxicity. Fill the outer wells with 200 µL of pyrogen-free water to prevent evaporative loss.
11. Leave cells to form a confluent monolayer by incubating overnight in a humidified 37°C incubator with 5%CO_2/95% air.

3.1.2. Preparation of Standards and Samples

1. Prepare TNF standard solutions: Dilute stock TNF to 833 pg/mL in tissue-culture medium. Prepare eight doubling dilutions of this standard, i.e., the lowest concentration being 3.25 pg/mL. This is most conveniently performed by using 1 mL of tissue-culture medium in eight sterile bijou tubes, and serially diluting 1 mL of the top standard in each of these tubes.
2. Dilute stock actinomycin solution to 8 µg/mL in tissue-culture medium. Add 25 µL of this solution to each well of the tissue-culture plate that contains cells. Do not remove the tissue-culture fluid that is already present.
3. Prepare the samples: Make suitable dilutions of samples in tissue-culture medium so that estimated TNF concentrations will fall within the standard curve (*see* **Note 3**).
4. Add 75 µL of each TNF standard in triplicate to the plate. Include a zero TNF well, i.e., three wells with 75 µL of tissue-culture medium.
5. Similarly add 75 µL of each sample to the tissue-culture plate. Assay in triplicate.
6. Incubate at 37°C as before for 20–24 h (*see* **Note 4**).

3.1.3. Reading the Bioassay Plate

1. After incubation, discard medium by inverting plate over sink.
2. Wash plate with 200 µL of RPMI per well, most conveniently added using a

multichannel pipet device. Remove medium by inversion as before, and remove excess fluid by rapping the inverted plate on a stack of tissues until dry.

3. Add 50 µL of formyl saline to each well to fix the cells. Incubate at room temperature for 10 min.
4. Discard the fixative (care, it is corrosive and toxic by inhalation).
5. Add 50 µL of crystal violet to each well and incubate for 10 min at room temperature.
6. Rinse with distilled water added from a wash bottle and tap dry on tissues as before.
7. Read the absorbance of each well using a plate reader at 540 or 580 nm. Be sure to read the absorbance of the outer wells that contained no cells (*see* **Note 5**).

3.1.4. Calculation of TNF Concentrations of Unknowns

1. Calculate the mean OD of the wells that contained no cells. Let this be A.
2. Calculate the mean OD of the standards and samples. Subtract A from all these values. Use these adjusted values in the formula below. The OD of the well with no TNF is termed $OD_{zero}TNF$. The OD of the wells with standards or unknowns are termed OD_{test}.
3. Calculate percentage cytotoxicity as follows:

$$\%\text{cytotoxicty} = 100 \times \left(OD_{zero}TNF - \frac{ODtest}{OZ_{zero}TNF}\right)$$

4. Plot percentage cytotoxicity against the $\log_{10}$ of the TNF standard concentration. This usually gives a sigmoid shaped curve (*see* **Fig. 1**). Read the unknown concentration of TNF by using its observed cytotoxicity (*see* **Note 6**).

3.2. TNF-β Bioassay

The bioassay for TNF-β is carried out exactly as described for TNF-α above. For the observed cytotoxicity to reflect TNF-β alone, neutralizing antibody to TNF-α must be added to the samples. The exact amount that must be added is dependent on the expected concentration of TNF-α that is present in the samples. The maker of the antiserum will quote the titer of the lot supplied in terms of the amount of TNF that a given volume of the antiserum will neutralize. The amount required can then be added to the samples, such that all the TNF-α present in the samples will be neutralized (*see* **Note 7**).

4. Notes

1. There are many different suppliers of recombinant TNF-α that can be used as a standard. It is important to realize that the specific activity of the cytokine will vary from supplier to supplier and even from lot to lot from the same source. Thus the same concentration of TNF in terms of a weight per unit volume will produce different cytotoxic effects if its specific activity varies. In practice, if

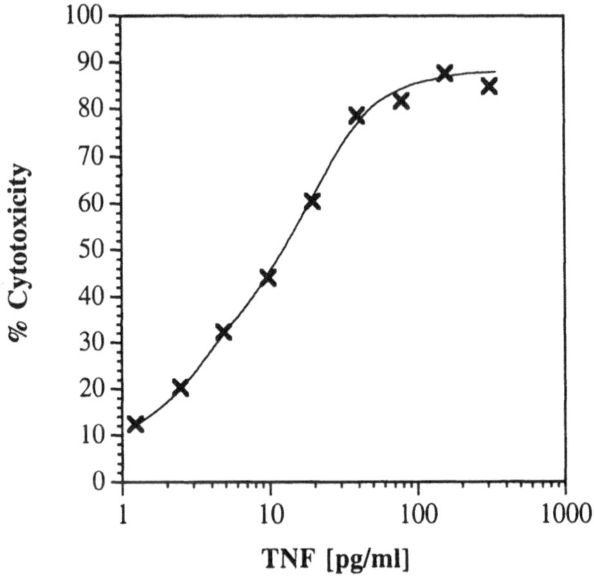

Fig. 1. Percent cytotoxicity in L929 cells treated with the indicated concentrations of murine TNF-α.

TNF is obtained from the indicated source and stored as described there is actually very little variation between lots. However, we have found that there can be quite large differences in specific activity of TNF from different suppliers. Thus, if a different supplier is used, the concentrations of TNF obtained using the new standard can be rather different from those obtained previously. It is obviously therefore important to try and keep to the same manufacturer as much as possible, and certainly within any one group of experiments.

Generally the supplier gives detailed information on the storage of the cytokine. It is important to keep the cytokine as concentrated as possible to avoid losses by absorbtion to the walls of the storage vessel—carrier protein such as bovine serum albumin (BSA) can be added if the cytokine must be diluted. Avoid freeze thaw cycles by freezing small aliquots and never refreezing a sample. Storage at –80°C gives much superior stability compared to storage at –20°C.

2. The quality of the L929 cells is the biggest single factor in determining the success and reproducibility of the TNF bioassay. The cells must be passaged in a vigorous growth phase. If they are allowed to overgrow it can take up to 1 wk before they recover fully. Under phase-contrast microscopy they should look even and with a darkish cytoplasm. If they are healthy they grow very fast, easily doubling in number with an overnight incubation. If the cells are not healthy at the beginning of the assay, it is not worth continuing.
3. The concentrations of TNF in the standard curve are those we have found to give a good range of cytotoxicity (**Fig. 1**). This may vary depending on the growth

state of the cells, etc., so that the concentrations might need to be modified by the experimenter to give a good range of cytotoxicity. We routinely are able to detect 1 pg/mL of TNF-α as the final concentration on the plate. This corresponds to 2.66 pg/mL in the added sample before dilution. Some thought should be given to the biological fluid that forms the samples. Very high concentrations of human serum might give rise to spurious cytotoxicity. Controls with zero TNF in the sample fluid to be tested should be performed, as well as a number of different dilutions of samples to check that there is cytotoxicity independent of the presence of human serum.

4. Cytotoxicity can be very rapid in this assay, appearing after 6 h or so. However, the plates are best left for at least 20 h to ensure that the lowest concentrations of TNF have produced a cytotoxic effect. In general, we have found the L929 assay to be free from interference from other substances, such as LPS or other cytokines. However, with the availability of neutralizing antibodies to TNF-α, it is always a good idea to check that the observed bioactivity can be absorbed with a neutralizing anti-TNF-α antibody. Such serum is available from Genzyme and should be added directly to a sample at a suitable concentration to absorb all the expected TNF-α. This need only be done on a sample with the highest measured TNF-α activity to check that it is all caused by TNF-α bioactivity. If there is residual cytotoxicity after addition of this antiserum, then this may also be caused by the presence of TNF-β. This can be checked by using a neutralizing serum to TNF-β in the same fashion.
5. An occasional problem with the crystal violet staining is patchy staining that makes the OD measurements unreliable. It is not clear why this happens from time to time. Either the assay must be repeated or the stain can be solubilized by the addition of 200 μL of 1% SDS to every well. The OD can then be read of the resulting solution, which will reflect the total amount of stain present in each well and hence degree of cytotoxicity.
6. It is important to perform the calculations as indicated. We usually end up with a sigmoid plot of $\log_{10}$ [TNF] vs percentage cytotoxicity (**Fig. 1**). Unknown TNF concentrations can either be read directly from this graph, or if available, a curve-fitting program can be used to generate an equation linking concentration of TNF to percent cytotoxicity. In any event, never derive a TNF concentration from a percent cytotoxicity reading that falls outside the range of the standard curve. Make a suitable adjustment to the concentration of the sample and re-assay.
7. In general, serum concentrations of TNF-β are much lower than TNF-α. However, it may be a significant component of serum from patients with Gram-positive sepsis and from in vitro experiments using stimulated lymphocytes.

References

1. Oettgen, H. F., Carswell, E. A., Kassel, R. L., Fiore, N., Williamson, B., Hoffmann, M. K., Haranaka, K., and and Old, L. J. (1980) Endotoxin-induced tumor necrosis factor. *Recent Results Cancer Res.* **75,** 207–212.

2. Tracey, K. J., Beutler, B., Lowry, S. F., Merryweather, J., Wolpe, S., Milsark, I. W., Hariri, R. J., Fahey, T. J. I., Zentella, A., Albert, J. D., Shires, G. T., and Cerami, A. (1986) Shock and tissue injury induced by recombinant human cachectin. *Science* **234,** 470–474.
3. Flick, D. A. and Gifford, G. E. (1984) Comparison of in vitro cell cytotoxic assays for tumor necrosis factor. *J. Immunol. Methods* **68,** 167–174.
4. Angehrn, P., Banner, D., Braun, T., d'Arcy, A., Gehr, G., Gentz, R., Mackay, F., Schlaeger, E.-J., Schoenfeld, H., Loetscher, H., and Lesslauer, W. (1993) Two distinct tumor necrosis factor receptors in health and disease, in *Tumor Necrosis Factor: Molecular and Cellular Biology and Clinical Relevance* (Fiers, W. and Buurman, W. A., eds.), Karger, Basel, Switzerland, pp. 33–39.
5. Waage, A., Halstensen, A., and Espevik, T. (1987) Association between tumor necrosis factor in serum and fatal outcome in patients with meningococcal disease. *Lancet* **i,** 355–357.
6. Gray, P. W., Aggarwal, B. B., Benton, C. V., Bringman, T. S., Henzel, W. J., Jarrett, J. A., Leung, D. W., Moffat, B., Ng, P., Svedersky, L. P., Palladino, M. A., and Nedwin, G. E. (1984) Cloning and expression of cDNA for human lymphotoxin, a lymphokine with tumour necrosis activity. *Nature* **312,** 721–724.
7. Sriskandan, S., Moyes, D., and Cohen, J. (1996) Detection of circulating bacterial superantigen and lymphotoxin-alpha in patients with streptococcal toxic-shock syndrome [letter]. *Lancet* **348,** 1315–1316.
8. Wong, G. H., Elwell, J. H., Oberley, L. W., and Goeddel, D. V. (1989) Manganous superoxide dismutase is essential for cellular resistance to cytotoxicity of tumor necrosis factor. *Cell* **58,** 923–931.
9. Pfister, H., Hennet, T., and Jungi, T. W. (1992) Lipopolysaccharide synergizes with tumour necrosis factor-alpha in cytotoxicity assays. *Immunology* **77,** 473–476.
10. Branch, D. R., Shah, A., and Guilbert, L. J. (1991) A specific and reliable bioassay for the detection of femtomolar levels of human and murine tumor necrosis factors. *J. Immunol. Methods* **143,** 251–261.

9

Assay of Soluble Tumor Necrosis Factor Receptors

Maarten G. Bouma and Wim A. Buurman

1. Introduction

Over the last decade, numerous basic biological as well as experimental and clinical studies have firmly established the significance of tumor necrosis factor (TNF) as a principal proximal mediator of sepsis *(1–4)*. One of the major insights that has emerged during recent years has been that under physiological circumstances, TNF activity is tightly controlled and locally restricted. In this respect, the soluble TNF receptors (sTNF-Rs) have been recognized to exert an important regulatory control on the biological actions of TNF, not only in the normal host defense against infection, but also in systemic inflammatory disorders that are related to infectious as well as noninfectious etiologies. Moreover, elevated systemic levels of sTNF-R have been demonstrated to have accurate diagnostic as well as prognostic significance in clinical sepsis and other critical illnesses *(5–13)*. Therefore, accurate determination of sTNF-R in plasma or serum has become an important tool to gain information about a variety of pathological conditions that are characterized by TNF-mediated immune activation, both in the experimental and in the clinical setting.

In this chapter we describe in detail an enzyme-linked immunosorbent assay (ELISA) for the quantification of human sTNF-R55 and sTNF-R75 in plasma. This ELISA is based on the monoclonal antibodies (MAbs) MR1-1 (anti-sTNF-R55) and MR2-2 (anti-sTNF-R75) as catching antibodies, respectively, and biotinylated polyclonal rabbit IgG directed against sTNF-R55 and sTNF-R75, respectively, as detection antibody. The captured antigen is quantified subsequently by a reaction with streptavidin-peroxidase conjugate. The assay is feasible for determination of human sTNF-R in plasma as well as in culture medium. In addition, a similar ELISA for determination of both sTNF-R in the

From: *Methods in Molecular Medicine, Vol. 36: Septic Shock*
Edited by: T. J. Evans © Humana Press Inc., Totowa, NJ

mouse is described. This ELISA, however, uses rat MAbs to mouse sTNF-R as capture antibodies, and biotinylated rabbit polyclonal antibodies for detection. Finally, we discuss some general issues of importance to keep in mind when using these techniques.

2. Materials

2.1. Human sTNF-R ELISA

1. Phosphate-buffered saline (PBS): Dissolve 4 g KCl, 4 g KH_2PO_4, 160 g NaCl, and 58 g $Na_2HPO_4.12H_2O$ in 2 L aquadest, autoclave, and store under sterile conditions. Dilute 1:10 in aquadest before use, pH should be 7.2–7.4.
2. Streptavidin-peroxidase conjugate (Zymed, San Francisco, CA).
3. 3,3′,5,5′-tetramethylbenzidine (TMB): Make fresh as required by mixing equal volumes of TMB substrate (Kirkegaard & Perry Laboratory, Gaithersburg, MD, cat. no. 50-76-00) and peroxidase solution B (cat. no. 50-76-03). Do not store the mixed solution.
4. 1 *M* H_2SO_4: Add 50 mL H_2SO_4 100% to 950 mL double-distilled water.
5. Anti-TNF-R55 MAb MR1-1, store at 4°C. Store only in polypropylene tubes (*see* **Note 1**).
6. Anti-TNF-R75 MAb MR2-2, store at 4°C. Store only in polypropylene tubes (*see* **Note 1**).
7. Soluble TNF-R55: Purified culture supernatant from NSO10 cells (*see* **Note 1**).
8. Soluble TNF-R75: Purified culture supernatant from NSO23 cells (*see* **Note 1**).
9. Biotinylated anti-TNF-R55 polyclonal rabbit IgG, light-sensitive, store at 4°C. Store only in polypropylene tubes (*see* **Note 2**).
10. Biotinylated anti-TNF-R75 polyclonal rabbit IgG, light-sensitive, store at 4°C. Store only in polypropylene tubes (*see* **Note 2**).
11. Wash buffer: 0.1% (v/v) Tween 20 (Serva, Heidelberg, Germany, cat. no. 37470) in double-distilled water.
12. Dilution buffer: 0.1% (v/v) BSA (Sigma, St. Louis, MO, cat. no. A-7906) in PBS.
13. Maxisorb F96 immunoplates (Nunc, Roskilde, Denmark, cat. no. 4-39454).

2.2. Murine sTNF-R ELISA

1. **Subheading 2.1., items 1–4**.
2. Anti-sTNF-R55 MAb HM103, store at 4°C. Store only in polypropylene tubes (*see* **Note 3**).
3. Anti-sTNF-R75 MAb HM102, store at 4°C. Store only in polypropylene tubes (*see* **Note 3**).
4. Soluble TNF-R55: purified culture supernatant from Bacculo cultures.
5. Soluble TNF-R75: purified culture supernatant from Bacculo cultures.
6. Biotinylated anti-murine sTNF-R55 polyclonal rabbit IgG, light-sensitive, store at 4°C. Store only in polypropylene tubes (*see* **Note 4**).
7. Biotinylated anti-murine sTNF-R75 polyclonal rabbit IgG, light-sensitive, store at 4°C. Store only in polypropylene tubes (*see* **Note 4**).

8. Wash buffer and dilution buffer (**Subheading 2.1., items 11** and **12**).
9. Maxisorb F96 immunoplates (Nunc cat. no. 4-39454).

3. Methods

3.1. Human sTNF-R ELISA

1. Coat all 96 wells of the ELISA plate with MAb MR1-1 (anti-sTNF-R55) or MR2-2 (anti-sTNF-R75) at approx 3 µg/mL (concentration according to the batch, *see* **Note 5**) in PBS, 100 µL/well. Incubate overnight at 4°C.
2. Empty the plate. To block nonspecific binding sites, add 150 µl/well PBS-1% BSA to all wells, and incubate 1 h at room temperature.
3. Prepare a standard titration curve of sTNF-R55 or sTNF-R75, respectively, by serial dilution 1:2 in PBS-0.1% BSA in polypropylene tubes, ranging from 5 ng/ml to 2.5 pg/ml.
4. Wash the plate five times with wash buffer. It is absolutely essential to leave the wash buffer on the plate for at least 1 min during each wash step.
5. Empty the plate. Add the prepared standard serial dilution of the sTNF-R to be assayed in duplicate to rows 1–3, 100 µL/well. Apply the experimental samples to rows 4–12 at 100 µL/well. For determination of background absorbance, leave at least eight wells without sample, only add control plasma or culture medium as appropriate, 100 µL/well. Incubate 2 h at room temperature (*see* **Note 6**).
6. Wash the plate five times with wash buffer.
7. Empty the plate. Add the appropriate biotinylated polyclonal rabbit anti-sTNF-R appropriately diluted in PBS-0.1% BSA (dilution according to the batch, *see* **Note 5**) to all wells, 100 µl/well. Incubate 1 h at room temperature.
8. Wash the plate five times with wash buffer.
9. Empty the plate. Add streptavidin-peroxidase conjugate diluted approx 1:5000 in PBS-0.1% BSA (dilution according to the batch, *see* **Note 5**) to all wells, 100 µL/well. Incubate 1 h at room temperature. Ascertain that the dilution buffer does not contain any azide.
10. Wash the plate five times with wash buffer.
11. Empty the plate. Add the freshly prepared TMB solution as substrate to all wells, 100 µL/well. Incubate 15 min at room temperature.
12. Stop the reaction by adding 100 µL/well 1 *M* H_2SO_4 to all wells.
13. Measure absorption at 450 nm using a micro ELISA autoreader. Use air as the blank reading.

3.2. Murine sTNF-R ELISA

1. Coat all 96 wells of the ELISA plate with the specific rat anti-mouse sTNF-R monoclonal at approx 3 µg/mL (concentration according to the batch, *see* **Note 5**) in PBS, 50 µL/well. Incubate overnight at 4°C.
2. Empty the plate. To block nonspecific binding sites, add 100 µL/well PBS-1% BSA to all wells, and incubate 1 h at room temperature.

3. Prepare a standard titration curve of recombinant murine sTNF-R55 ranging from 5 ng/mL to 2.5 pg/mL, or of recombinant murine sTNF-R75 ranging from 50 ng/mL to 25 pg/mL, by serial dilution 1:2 in PBS-0.1% BSA in polypropylene tubes.
4. Wash the plate five times with wash buffer. It is absolutely essential to leave the wash buffer on the plate for at least 1 min during each wash.
5. Empty the plate. Add the prepared standard serial dilution of the murine sTNF-R to be assayed in duplicate to rows 1–3, 50 µL/well. Apply the experimental samples to rows 4–12 at 50 µL/well. Leave at least eight wells without sample, only add PBS-0.1% BSA, 50 µL/well. Incubate 1 h at room temperature.
6. Wash the plate five times with wash buffer.
7. Empty the plate. Add the appropriate biotinylated polyclonal rabbit anti-sTNF-R diluted approx 1:3000 in PBS-0.1% BSA (dilution according to the batch, *see* **Note 5**) to all wells, 50 µL/well. Incubate 1 h at room temperature.
8. Wash the plate five times with wash buffer.
9. Empty the plate. Add streptavidin-peroxidase conjugate diluted approx 1:5000 in PBS-0.1% BSA (dilution according to the batch, *see* **Note 5**) to all wells, 50 µL/well. Incubate 1 h at room temperature. Ascertain that the dilution buffer does not contain any azide.
10. Wash the plate five times with wash buffer.
11. Empty the plate. Add the freshly prepared TMB solution as substrate to all wells, 50 µL/well. Incubate 15 min at room temperature.
12. Stop the reaction by adding 50 µL/well 1 *M* H_2SO_4 to all wells.
13. Measure absorption at 450 nm using a micro ELISA autoreader.

4. Notes

1. The MAbs specifically directed against sTNF-R55 and sTNF-R75 were obtained by immunizing BALB/c mice with sTNF-R55 and sTNF-R75, respectively, which were purified from the culture supernatant of NSO cells transfected with the extracellular part of either TNF-R55 or TNF-R75 (NSO10 cells and NSO23 cells, respectively, provided by Celltech, (Slough, UK), by affinity chromatography on a TNF-coupled agarose column (Amino Link column, Pierce, Rockford, IL). Subsequently, spleen cells of immunized mice were fused with Sp2/0 mouse myeloma cells, and growing hybridomas were tested for production of MAb specifically reactive with sTNF-R55 or sTNF-R75. Monoclonal antibody MR1-1 (anti-sTNF-R55) and MAb MR2-2 (anti-sTNF-R75) are both of the murine IgG_1 isotype.
2. Polyclonal rabbit antisera to sTNF-R55 and sTNF-R75 were obtained by immunizing rabbits with the purified sTNF-R55 and sTNF-R75, respectively. These antisera were subsequently biotinylated using standard techniques.
3. The MAbs directed specifically against murine sTNF-R55 and sTNF-R75 were obtained by immunizing Wistar rats with highly purified murine sTNF-R55 and sTNF-R75, respectively, which were purified from Bacculo culture supernatant by affinity chromatography on a murine TNF-coupled agarose column (Amino

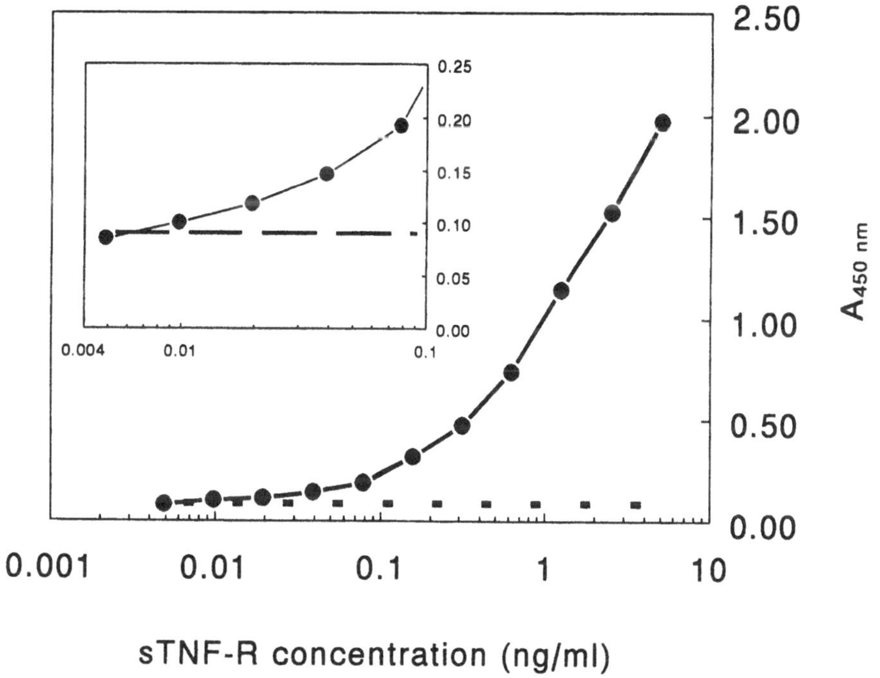

Fig. 1. Typical standard titration curve of sTNF-R. Values represent the mean of duplicate determinations. Mean background absorbance plus three times the SD is indicated by a dashed line.

Link column, Pierce, Rockford, IL). Subsequently, spleen cells of immunized rats were fused with Sp2/0 mouse myeloma cells, and growing hybridomas were tested for production of MAb specifically reactive with murine sTNF-R55 or sTNF-R75. Monoclonal antibody HM103 (anti-sTNF-R55) and MAb HM102 (anti-sTNF-R75) are both of the rat IgG_{2a} isotype.

4. Polyclonal rabbit antisera to sTNF-R55 and sTNF-R75 were obtained by immunizing rabbits with the purified murine sTNF-R55 and sTNF-R75, respectively. These antisera were subsequently biotinylated using standard techniques.
5. Each batch of monoclonal and polyclonal antibody, as well as each batch of streptavidin-peroxidase conjugate should be pretested to determine the optimal concentrations for use in the ELISA.
6. A typical standard sTNF-R titration curve, with values representing the mean absorbance of duplicate determinations, is depicted in **Fig. 1**. From this curve the sensitivity of the ELISA, which may be defined arbitrarily as the lowest detectable sTNF-R concentration giving rise to an absorbance higher than the mean background absorbance plus three times the standard deviation (SD) of the background, can be determined. The sensitivity of the ELISA, defined as such, is approx 10 pg/mL for both human sTNF-Rs.

7. The determination of sTNF-R may have certain advantages over TNF measurement in monitoring inflammatory states and predicting clinical outcome. Compared to circulating TNF, which is only shortly present and can therefore be easily missed, sTNF-R are detectable for a more prolonged period of time. In addition, sTNF-R are very stable, and their determination seems to be less prone to artifacts than that of TNF *(14)*. Also, the presence of sTNF-R may have a masking effect on measurement of TNF, depending on the specific TNF assay used *(15)*, which can easily lead to misinterpretation of results and may severely hamper cross-laboratory comparison of data concerning TNF levels. On the other hand however, a significant technical complication of sTNF-R measurement is formed by the presence of very high levels of TNF in the samples, such as in plasma of patients treated with TNF as antitumor treatment, e.g., regional TNF perfusion. Such extremely high concentrations of TNF (1000-fold or higher than seen in sepsis patients) can interfere with the interaction of the antibodies used in the ELISA with sTNF-R. Therefore, when determining sTNF-R in such samples, care should be taken to exclude the influence of TNF on the detection of sTNF-R. In our institute we use a special set of reagents for such measurements. Finally, whereas measurement of cytokines in plasma is hampered by rigid protocols to process the blood after sampling to prevent either cytokine release after sampling or loss of cytokine content during treatment and storage of the sample, the assay of sTNF-R does not demand special precautions in this respect. We do, however, recommend storing samples no longer than 30 min at room temperature as neutrophils can shed TNF-R very rapidly *(16,17)*.

   The determination of sTNF-R as accurate markers of inflammation has proven to be of value for many fields of medical research, not only for studies on sepsis *5,9–11,18,19)* and various infectious diseases *(20,21)*, including human immunodeficiency virus (HIV) *(22–24)*, but also for experimental and clinical investigations of a variety of other clinical entities, ranging from cancer *(25–27)*, atherosclerosis *(28,29)*, and heart failure *(30)*, to cardiac bypass surgery *(10,13)*, and trauma *(6,31)*. Moreover, quantification of sTNF-R as mediators of inflammation has provided a link to the metabolic derangements associated with several of these clinical disorders *(32–36)*. However, although sTNF-R are considered generally to be reliable diagnostic and prognostic indicators of inflammation, it is of importance to realize that sTNF-R are presumed to be part of anti-inflammatory mechanisms as well. Whereas the assumed property of sTNF-R to stabilize the trimeric structure of TNF in the circulation *(37)*, thereby prolonging TNF bioactivity, can be regarded as pro-inflammatory, their capacity to inactivate *(38,39)* and possibly enhance clearance of circulating TNF *(1,40,41)* can be considered anti-inflammatory. Thus, during sepsis the increased presence of sTNF-R in a large molar excess over TNF may function as an endogenous inhibitory and clearance mechanism of TNF bioactivity, which in severe cases may still be insufficient to prevent the lethal consequences of high systemic TNF concentrations *(18)*.

   Finally, with regard to the interpretation of enhanced sTNF-R levels there is one important caveat to be remembered. As sTNF-R are mainly cleared by the kidney, renal impairment may considerably influence systemic sTNF-R concentrations *(41)*. Consistent with the central role of the kidney in sTNF-R clearance are the observations of enhanced sTNF-R levels positively correlating with plasma creatinine levels in patients with acute or chronic renal failure *(42,43)*, as well as in patients with the sepsis syndrome *(9)*. Thus, the presence of renal failure may result in an unproportional elevation of circulating sTNF-R, that does not directly reflect the inflammatory state, but rather the degree of renal failure.

References

1. Bemelmans, M. H. A., Van Tits, L. J. H., and Buurman, W. A. (1996) Tumor necrosis factor: function, release and clearance. *Crit. Rev. Immunol.* **16,** 1–11.
2. Strieter, R. M., Kunkel, S. L., and Bone, R. C. (1993) Role of tumor necrosis factor-α in disease states and inflammation. *Crit. Care Med.* **21,** S447–S463.
3. Tracey, K. J. and Cerami, A. (1993) Tumor necrosis factor: an updated review of its biology. *Crit. Care Med.* **21,** S415–S422.
4. Beutler, B. and Grau, G. E. (1993) Tumor necrosis factor in the pathogenesis of infectious diseases. *Crit. Care Med.* **21,** S423–S435.
5. Girardin, E., Roux-Lombard, P., Grau, G. E., Suter, P., Galati, H., and Dayer, J. M. (1992) Imbalance between tumour necrosis factor-α and soluble TNF receptor concentrations in severe meningococcocaemia. *Immunology* **76,** 20–23.
6. Cinat, M. E., Waxman, K., Granger, G. A., Pearce, W., Annas, C., and Daughters, K. (1994) Trauma causes sustained elevation of soluble tumor necrosis factor receptors. *J. Am. Coll. Surg.* **179,** 529–537.
7. Rogy, M. A., Coyle, S. M., Oldenburg, H. S., Rock, C. S., Barie, P. S., Van Zee, K. J., Smith, C. G., Moldawer, L. L., and Lowry, S. F. (1994) Persistently elevated soluble tumor necrosis factor receptor and interleukin-1 receptor antagonist levels in critically ill patients. *J. Am. Coll. Surg.* **178,** 132–138.
8. Kaufmann, P., Tilz, G. P., Lueger, A., and Demel, U. (1997) Elevated plasma levels of soluble tumor necrosis factor receptor (sTNFRp60) reflect severity of acute pancreatitis. *Intensive Care Med.* **23,** 841–848.
9. Froon, A. H. M., Bemelmans, M. H. A., Greve, J. W., Van der Linden, C. J., and Buurman, W. A. (1994) Increased plasma concentrations of soluble tumor necrosis factor receptors in sepsis syndrome: correlation with plasma creatinine values. *Crit. Care Med.* **22,** 803–809.
10. Pilz, G., Fraunberger, P., Appel, R., Kreuzer, E., Werdan, K., Walli, A., and Seidel, D. (1996) Early prediction of outcome in score-identified, postcardiac surgical patients at high risk for sepsis, using soluble tumor necrosis factor receptor-p55 concentrations. *Crit. Care Med.* **24,** 596–600.
11. Van Deuren, M., Van der Ven-Jongekrijg, J., Demacker, P. M. N., Burtelink, A. K. M., Van Dalen, R., Sauerwein, R. W., Gallati, H., Vannice, J. L., and Vander Meer, J. M. M. (1994) Differential expression of proinflammatory cytokines and their inhibitors during the course of meningococcal infections. *J. Infect. Dis.* **169,** 157–161.

12. De Beaux, A. C., Goldie, A. S., Ross, J. A., Carter, D. C., and Fearon, K. C. (1996) Serum concentrations of inflammatory mediators related to organ failure in patients with acute pancreatitis. *Br. J. Surg.* **83,** 349–353.
13. Maessen, J. G., Fransen, E. J., Dentener, M. A., Gorgels, A. P. M., and Buurman, W. A. (1996) Serum soluble TNF receptor as a risk factor in patients with heart failure having cardiac surgery. *J. Mol. Cell. Cardiol.* **28,** Abg (abstract).
14. Engelberts, I., Möller, A., Schoen, G. J. M., Van der Linden, C. J., and Buurman, W. A. (1991) Evaluation of measurement of human TNF in plasma by ELISA. *Lymphokine Cytokine Res.* **10,** 69–76.
15. Engelberts, I., Stephens, S., Francot, G. J. M., Van der Linden, C. J., and Buurman, W. A. (1991) Evidence for different effects of soluble TNF-receptors on various TNF measurements in human biological fluids. *Lancet* **338,** 515–516.
16. Porteu, F. and Nathan, C. (1990) Shedding of tumor necrosis factor receptors by activated human neutrophils. *J. Exp. Med.* **172,** 599–607.
17. Lantz, M., Björnberg, F., Olsson, I., and Richter, J. (1994) Adherence of neutrophils induces release of soluble tumor necrosis factor receptor forms. *J. Immunol.* **152,** 1362–1369.
18. Van der Poll, T., Jansen, J., Van Leenen, D., Von der Möhlen, M., Levi, M., Ten Cate, H., Gallati, H., Ten Cate, J. W., and Van Deventer, S. J. (1993) Release of soluble receptors for tumor necrosis factor in clinical sepsis and experimental endotoxemia. *J. Infect. Dis.* **168,** 955–960.
19. Olszyna, D. P., Prins, J. M., Buis, B., Van Deventer, S. J., Speelman, P., and Van der Poll, T. (1998) Levels of inhibitors of tumor necrosis factor alpha and interleukin 1beta in urine and sera of patients with urosepsis. *Infect. Immun.* **66,** 3527–3534.
20. Mwatha, J. K., Kimani, G., Kamau, T., Mbugua, G. G., Ouma, J. H., Mumo, J., Fulford, A. J., Jones, F. M., Butterworth, A. E., Roberts, M. B., and Dunne, D. W. (1998) High levels of TNF, soluble TNF receptors, soluble ICAM-1, and IFN-gamma, but low levels of IL-5, are associated with hepatosplenic disease in human schistosomiasis mansoni. *J. Immunol.* **160,** 1992–1999.
21. Bethell, D. B., Flobbe, K., Cao, X. T., Day, N. P., Pham, T. P., Buurman, W. A., Cardosa, M. J., White, N. J., and Kwiatkowski, D. (1998) Pathophysiologic and prognostic role of cytokines in dengue hemorrhagic fever. *J. Infect. Dis.* **177,** 778–782.
22. Godfried, M. H., Romijn, J. A., Van der Poll, T., Weverling, G. J., Corssmit, E. P., Endert, E., Eeftinck-Schattenkerk, J. K., and Sauerwein, H. P. (1995) Soluble receptors for tumor necrosis factor are markers for clinical course but not for major metabolic changes in human immunodeficiency virus infection. *Metabolism* **44,** 1564–1569.
23. Hober, D., Benyoucef, S., Delannoy, A. S., De Groote, D., Ajana, F., Mouton, Y., and Wattre, P. (1996) Plasma levels of sTNFR p75 and IL-8 in patients with HIV-1 infection. *Immunol. Lett.* **52,** 57–60 .
24. Stein, D. S., Lyles, R. H., Graham, N. M., Tassoni, C. J., Margolick, J. B., Phair, J. P., Rinaldo, C., Detels, R., Saah, A., and Bilello, J. (1997) Predicting clinical

progression or death in subjects with early-stage human immunodeficiency virus (HIV) infection: a comparative analysis of quantification of HIV RNA, soluble tumor necrosis factor type II receptors, neopterin, and beta2-microglobulin. Multicenter AIDS Cohort Study. *J. Infect. Dis.* **176,** 1161–1167.

25. Aderka, D., Engelmann, H., Hornik, V., Skornick, Y., Levo, Y., Wallach, D., and Kushtai, G. (1991) Increased serum levels of soluble receptors for tumor necrosis factor in cancer patients. *Cancer Res.* **51,** 5602–5607.
26. Gadducci, A., Ferdeghini, M., Fanucchi, A., Annicchiarico, C., Prato, B., Prontera, C., Facchini, V., and Genazzani, A. R. (1996) Serum levels of soluble receptors for tumor necrosis factor (p55 and p75 sTNFr) in endometrial cancer. *Anticancer Res.* **16,** 3125–3128.
27. Staal-Van den Brekel, A. J., Dentener, M. A., Drent, M., Ten Velde, G. P., Buurman, W. A., and Wouters, E. F. (1998) The enhanced inflammatory response in non-small cell lung carcinoma is not reflected in the alveolar compartment. *Respir. Med.* **92,** 76–83.
28. Elneihoum, A. M., Falke, P., Hedblad, B., Lindgarde, F., and Ohlsson, K. (1997) Leukocyte activation in atherosclerosis: correlation with risk factors. *Atherosclerosis* **131,** 79–84.
29. Elneihoum, A. M., Falke, P., Axelsson, L., Lundberg, E., Lindgarde, F., and Ohlsson, K. (1996) Leukocyte activation detected by increased plasma levels of inflammatory mediators in patients with ischemic cerebrovascular diseases. *Stroke* **27,** 1734–1738.
30. Anker, S. D., Clark, A. L., Kemp, M., Salsbury, C., Teixeira, M. M., Hellewell, P. G., and Coats, A. J. (1997) Tumor necrosis factor and steroid metabolism in chronic heart failure: possible relation to muscle wasting. *J. Am. Coll. Cardiol.* **30,** 997–1001.
31. Tan, L. R., Waxman, K., Scanell, G., Loli, G., and Granger, G. A. (1993) Trauma causes early release of soluble receptors for tumor necrosis factor. *J. Trauma* **34,** 634–638.
32. Suttmann, U., Ockenga, J., Schneider, H., Selberg, O., Schlesinger, A., Gallati, H., Wolfram, G., Deicher, H., and Muller, M. J. (1996) Weight gain and increased concentrations of receptor proteins for tumor necrosis factor after patients with symptomatic HIV infection received fortified nutrition support. *J. Am. Diet. Assoc.* **96,** 565–569.
33. Staal-Van den Brekel, A. J., Schols, A. M., Dentener, M. A., Ten Velde, G. P., Buurman, W. A., and Wouters, E. F. (1997) Metabolism in patients with small cell lung carcinoma compared with patients with non-small cell lung carcinoma and healthy controls. *Thorax* **52,** 338–341.
34. Schols, A. M., Buurman, W. A., Staal-Van den Brekel, A. J., Dentener, M. A., and Wouters, E. F. (1996) Evidence for a relation between metabolic derangements and increased levels of inflammatory mediators in a subgroup of patients with chronic obstructive pulmonary disease. *Thorax* **51,** 819–824.
35. Staal-Van den Brekel, A. J., Schols, A. M., Dentener, M. A., Ten Velde, G. P., Buurman, W. A., and Wouters, E. F. (1997) The effects of treatment with chemo-

therapy on energy metabolism and inflammatory mediators in small-cell lung carcinoma. *Br. J. Cancer* **76,** 1630–1635.

36. Staal-Van den Brekel, A. J., Dentener, M. A., Schols, A. M., Buurman, W. A., and Wouters, E. F. (1995) Increased resting energy expenditure and weight loss are related to a systemic inflammatory response in lung cancer patients. *J. Clin. Oncol.* **13,** 2600–2605.
37. Aderka, D., Engelmann, H., Maor, Y., Brakebusch, C., and Wallach, D. (1992) Stabilization of the bioactivity of tumor necrosis factor by its soluble receptors. *J. Exp. Med.* **175,** 323–329.
38. Bazzoni, F. and Beutler, B. (1996) The tumor necrosis factor ligand and receptor families. *N. Eng. J. Med.* **334,** 1717–1725.
39. Van Zee, K. J., Kohno, T., Fischer, E., Rock, C. S., Moldawer, L., and Lowry, S. F. (1992) Tumor necrosis factor soluble receptors circulate during experimental and clinical inflammation and can protect against excessive tumor necrosis factor-alpha in vitro and in vivo. *Proc. Natl. Acad. Sci. USA* **89,** 4845–4849.
40. Bemelmans, M. H. A., Gouma, D. J., and Buurman, W. A. (1994) Tissue distribution and clearance of soluble murine TNF receptors in mice. *Cytokine* **6,** 608–615.
41. Bemelmans, M. H. A., Gouma, D. J., and Buurman, W. A. (1993) Influence of nephrectomy on tumor necrosis factor clearance in a murine model. *J. Immunol.* **150,** 2007–2017.
42. Peetre, C., Thysell, H., Grubb, A., and Olsson, I. (1988) A tumor necrosis factor binding protein is present in human biological fluids. *Eur. J. Haematol.* **41,** 414–419.
43. Brockhaus, M., Bar-Khayim, Y., Gurwicz, S., Frensdorff, A., and Haran, N. (1992) Plasma tumor necrosis factor soluble receptors in chronic renal failure. *Kidney Int.* **42,** 663–667.

10

Whole-Blood Assays for Cytokine Production

Daniel G. Remick, David E. Newcomb, and Jon S. Friedland

1. Introduction

Substantial interest has been generated by the potential roles of the cytokines in health and disease *(1)*. This has prompted considerable investigation into how these mediators are regulated to answer such basic questions as which stimuli initiate transcription and what factors are responsible for inhibiting secretion. This has resulted in elegant studies that have begun to define the intracellular-signaling pathways responsible for the upregulation of cytokines *(2,3)*. Many of these studies have been done with cultures of cell lines derived from cancers or primary cultures of isolated fibroblasts, endothelial cells, or isolated mononuclear cells. Whereas these studies have provided substantial insight, they may be limited in their scope because they do not include all cell–cell or cell–protein interactions that take place in vivo. Other studies have used endotoxin injection into normal human volunteers to study the upregulation of cytokines *(4,5)*. These studies with the normal volunteers provide precise information about the kinetics of cytokine production, but they are difficult to perform and very expensive. The whole-blood model serves as a useful bridge between using normal volunteers and isolated peripheral blood mononuclear cells. For critically ill patients it would be impossible to perform endotoxin infusion studies, and it would even be difficult to conduct these types of studies in chronically ill patients. Whole blood may also be used to study the immune responses of such patients in an attempt to determine how their cytokine regulation differs from normal individuals.

The whole-blood model offers several advantages over traditional isolated cells. First, the cell isolation procedure requires some equipment and reagents that may not be available in remote locations. This makes the method widely

From: *Methods in Molecular Medicine, Vol. 36: Septic Shock*
Edited by: T. J. Evans © Humana Press Inc., Totowa, NJ

available even in regions in which serious diseases are present but sophisticated laboratories are not. Second, cell isolation may activate the cells resulting in high background levels of cytokines *(6)*. Third, the 100% plasma present in the whole blood is a much more physiologic culture medium than even the most comprehensive tissue culture formulation. Fourth, very small volumes of blood are required to set up the assay, and this can be very important for studies involving children *(7)*. It should be mentioned that there are drawbacks with the whole-blood system. As neutrophils represent the major nucleated cell in the blood there is a strong bias towards neutrophils. It is also more difficult to determine which cell is responsible for the production of which cytokine in the heterogeneous cell population, although this can be overcome by doing intracellular cytokine staining and flow cytometery or using specific cell stimulation.

1.1. Previous Work with Whole-Blood Stimulation Studies

Several hundred publications have documented the use of whole blood for the investigation of the immune response and it is beyond the scope of this chapter to review them all. However, we would like to briefly reference how this method has been used previously. Particularly relevant for this series are those studies that have examined the use of whole blood in sepsis. A study by Kremer et al. used lipopolysaccharide (LPS)-stimulated whole blood from patients with severe sepsis and measured the production of TNF, IL-1β and IL-6 *(8)*. The patients with severe sepsis had a significant reduction in the capacity of whole blood to produce the pro-inflammatory cytokines. Normal volunteers infused with endotoxin have a reduced capacity to produce the same pro-inflammatory cytokines *(9)*. In patients with infections from the intensive care unit, there was decreased production of TNF and IL-6 in response to endotoxin, but augmentation of these cytokines when the blood is stimulated with either *Salmonella typhimurium* or *Staphylococcus aureus* *(10)*. Whereas the whole-blood method has the most direct application to sepsis because the blood is the target organ, it has been used to investigate other sorts of infections including viruses and mycobacteria *(11)*.

Whole blood has also been used to study the immune function of patients. A group of patients at increased risk for developing sepsis, those with severe blunt trauma, also have reduced production of cytokines *(12)*. Several studies have documented reduced cytokine production in patients with malignancies. Patients with chronic lymphocytic leukemia *(13)*, small-cell carcinoma of the lung *(14)*, melanoma *(15)*, and colorectal carcinomas *(16)* all have reduced production of cytokines in stimulated whole blood compared to normal controls. In contrast, patients with a nonmalignant chronic disease (rheumatoid arthritis) have an increase in cytokine production *(17)*. Furthermore, the method may be used to follow-up changes in immune function. In a sequential study of patients with tuberculosis followed over 9 mo of treatment, persistently abnormal IL-8 secretion was detected despite apparent mycobacteriological cure *(11)*.

Cytokine production in whole blood has been studied extensively in patients with multiple sclerosis. In a novel use of the whole-blood system it has been shown to predict clinical relapses *(18)*. In a follow-up paper to this initial report, the original findings were confirmed and a peak of TNF production was observed to precede the relapse *(19)*. These findings were confirmed by another group which found increases of both TNF and IL-6 during relapses of multiple sclerosis *(20)*.

One note of caution needs to be raised about the whole-blood system. By design, this examines the peripheral blood and not the cells from the site of inflammation. Cells from the site of inflammation may act differently than the peripheral blood cells. In recent work with osteomyelitis there were no differences between patients and controls when the whole-blood pro-inflammatory secretion was measured *(21)*, but subsequent investigations have demonstrated that there are very high levels of pro-inflammatory cytokines at the actual site of infection.

1.2. Use of the Whole-Blood System to Study Drug Interactions

The whole-blood system has been used in several studies to examine drug interactions. Using this approach the same compounds may have different effects on different cytokines. For example, noradrenaline decreases TNF and IL-6 production *(22)*, whereas epinephrine enhances IL-8 production in the same system *(23)*. The whole-blood system has been used specifically to measure the ability of anti-inflammatory compounds to inhibit cytokine production and several of these have been shown to decrease the production of pro-inflammatory cytokines *(24,25)*.

One of the more intriguing aspects of the whole-blood system is its use compared to isolated mononuclear cells. Different effects may be observed depending on the experimental conditions. Using a synthetic lipid A analog as the stimulus, no TNF production was observed in the whole blood but isolated monocytes that were stimulated produced TNF *(26)*. Additional work will be required to determine which result is more biologically relevant.

1.3. Choice of Anticoagulants in Whole-Blood System

The choice of anticoagulants will have a direct impact on the results. These effects may result in false high or low values for the cytokines and have different effects depending on whether one is evaluating the baseline or the maximal stimulated production. Blood may be anticoagulated with ethylene diamine tetraacetic acid (EDTA) and still be stimulated to produce cytokines, but the values are lower than those observed when using heparin. The reason for the difference is multifactorial. First, EDTA will chelate the calcium and interfere with signal transduction, thereby decreasing the stimulus to the cells. This will result in lower background levels, and lower stimulated levels. The calcium effect may be mimicked by adding the calcium-channel inhibitor verapamil to heparin-anticoagulated blood. Some forms of heparin are contaminated with endotoxin which will result in some stimulation of the whole blood ***(27)***. This leads to higher background levels and higher stimulated levels. This problem may be resolved by filling the syringes with heparin that is used for injection into patients, rather than just the heparin anticoagulated tubes used for drawing blood. In our experience, the background production of cytokines is extremely low using high-quality heparin ***(28–32)***.

2. Materials

1. Heparin for injection (Elkins-Sinn Cherry Hill, NJ).
2. Centrifuge tubes (1.5-mL), either sterile or autoclavable (WST Promega, Madison, WI).
3. RPMI 1640 tissue-culture media (BioWhittaker, Walkersville, MD).
4. Lipopolysaccharide, *Escherichia coli* O111:B4 (Sigma, Poole, Dorset, UK).
5. Phytohemagglutinin (Sigma).
6. Zymosan A (Sigma).
7. Sterile, disposable Pasteur pipets.
8. Red-blood-cell lysing solution: EDTA tetra sodium salt 1 m*M*, $NaHCO_3$ 10 m*M*, NH_4Cl 150 m*M*.

3. Methods

The method is really simplicity itself. The short version of the method is to add anticoagulated blood to individual tubes with the stimulus, incubate at 37°C, and then collect the plasma.

3.1. Drawing Blood

1. Prepare heparinized syringes with 10 U/mL of heparin (*see* **Note 1**). Draw sufficient heparin to achieve the correct final concentration (10 U/mL).
2. Using sterile technique, draw the required amount of blood into the syringe. This blood may be obtained at the same time that blood is being drawn for routine analyses, such as electrolytes or white-blood-cell count.

3.2. Stimulating the Blood

1. If using an inhibitor, add this first to the bottom of the 1.5-mL centrifuge tube. Keep the total volume of the inhibitor to < 50 µL.
2. Add 1 mL of blood.

3. Add the appropriate stimulus, keeping the volume to < 50 μL (*see* **Note 2**). The choice of stimulus depends on the experiment design. In our hands we have observed excellent cytokine production with the following concentrations of stimuli: LPS 1 μg/mL; PHA 10 μg/mL; Zymosan, 100 μg/mL. Slightly higher levels of cytokines will be induced by higher concentrations of the stimuli, and lower levels by lower concentrations of stimuli. Some authors have mixed the blood 1:1 up to 1:10 with tissue culture media. We have not found that to be necessary.

3.3. Incubating the Blood

1. Place the blood at 37°C for the required length of time. For optimal stimulation the samples need to be rocked or agitated, and kept at 37°C. Many of the cytokines will reach peak production within the first 12 h (**Fig. 1A**). However IL-8 production continues to increase well beyond 24 h (**Fig. 1B**). Caution must be used with long incubation times as the blood will begin to hemolyze and cell viability is reduced. Despite the decrease in cell viability, IL-8 production continues to increase (*see* **Note 3**).

3.4. Collecting the Plasma

1. Centrifuge the blood at 600*g* for 5 min.
2. Remove the plasma and freeze, or run assays immediately. If the plasma is to be frozen, it is best kept at –80°C (*see* **Note 4**).

3.5. Working with the Leukocyte Pellet

After removal of the plasma it may be desirable to work with the white blood cells. To do this the red blood cells (RBC) must be removed. This is most conveniently done as follows:

1. Add the RBC lysing solution directly to the 1.5-mL centrifuge tube.
2. Allow the blood to incubate at room temperature for 5 min with end-over-end mixing.
3. Centrifuge the blood (600g for 5 min).
4. Aspirate the supernatant and add fresh RBC lysing solution to the pellet.
5. Gently resuspend the pellet and incubate for 5 min.
6. Centrifuge and aspirate the supernatant.
7. The cell pellet may now be used for extraction to measure mRNA, cell associated proteins, or the entire pellet may be used for cell viability assays.

3.6. Cell Viability

Cell viability may be easily measured in the cell pellet.

1. Resuspend the entire cell pellet in 1 mL of RPMI 1640 tissue-culture media.
2. Place triplicate, 100-μL aliquots of the resuspended cells in each well of a 96-well plate.

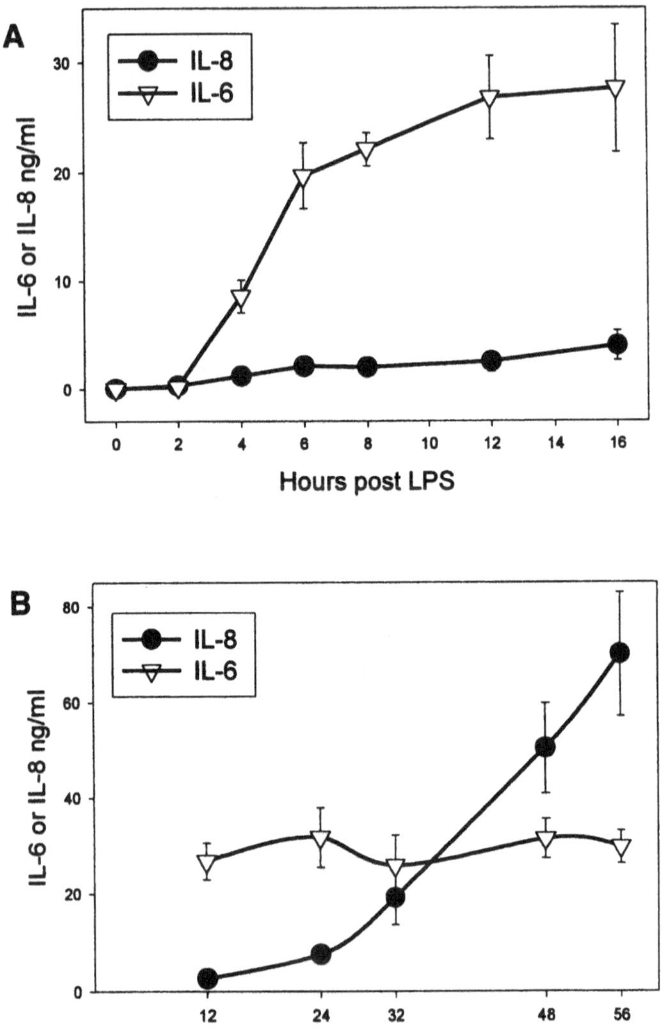

Fig. 1. Kinetics of cytokine production: Human whole blood was stimulated with 1 μg/mL of LPS and plasma collected at the indicated time points. LPS-stimulated IL-8 and IL-6 production (**A**) over the first 16 h, (**B**) from 12–56 h. IL-6 production has peaked by 12 h, but IL-8 production continues to increases. Each value is the mean ± SEM for five donors.

3. Dilute WST 1 : 1 with RPMI 1640, and add 5 μL/well to each well.
4. Incubate the plate overnight at 37°C and read on an ELISA reader after 16 h (read at 465 nm with a reference wavelength of 630 nm).

Table 1
Production of TNF or IL-8 in Human Whole Blood

TNF production			
Stimulus	Mean pg/mL	Minimum pg/mL	Maximum pg/mL
None	6	0	56
LPS 10 ng/mL	610	13	4400
PHA 1 µg/mL	410	85	1200
Zymosan A 100 µg/mL	5400	2900	10,100
IL-8 production			
Stimulus	Mean pg/mL	Minimum pg/mL	Maximum pg/mL
None	841	0	10,800
LPS 10 ng/mL	24,900	1400	132,000
PHA 1 µg/mL	54,800	2700	120,000
Zymosan A 100 µg/mL	125,000	2400	447,000

Human whole blood was anticoagulated with heparin and incubated for 24 h with the indicated stimuli. Plasma levels of TNF and IL-8 were determined 24 h later. These values are the combined data from 9–18 normal volunteers. This table provides the range that may be observed when using these stimuli.

3.7. Assays

Any assay that may be done with plasma may be done with the plasma from stimulated whole blood. Biologically active cytokines may be measured using bioassays or immunologically reactive cytokines may be measured.

4. Notes

1. The choice of anticogulant is critical. The same levels of cytokines will not be found in the same normal volunteer if different types of anticoagulants are used. Regardless of which anticoagulant is selected, it is necessary to use this same anticoagulant for all remaining studies in that experiment. We have observed a substantial increase in the production of pro-inflammatory cytokines when low-molecular-weight heparin was used rather than unfractionated heparin *(33)*. When using EDTA as the anticoagulant it is important to note that this may cause trouble in subsequent bioassays. Tissue-culture media contains calcium salts and as the plasma is diluted the EDTA concentration decreases while the calcium concentration remains the same. At some point in the dilutions there will be sufficient calcium to activate the clotting cascade. This problem may be overcome by including a 1 : 1000 dilution of heparin in the tissue-culture media.
2. Different stimuli will produce different levels of cytokines. Additionally, the same stimulus will produce different levels of different cytokines. For example, following an LPS stimulation there is substantially more IL-8 produced compared to TNF (*see* **Table 1**).

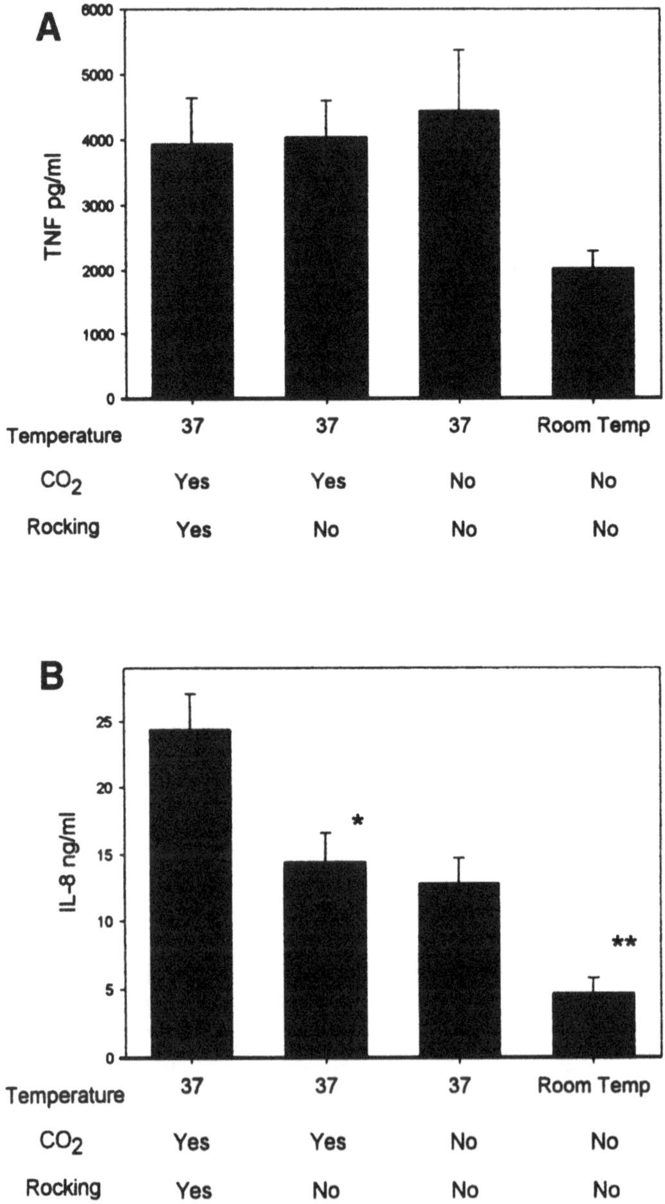

Fig. 2. Effects of different incubation conditions: Whole blood was incubated with 1 µg/mL of LPS in four different conditions as indicated. Plasma TNF and IL-8 was determined 24 h later. Only temperature affected TNF production while both rocking and temperature affected IL-8 production. Each value is the mean ± SEM for eight donors. (*) $p < 0.05$ compared to 37°C, plus rocking and CO_2, (**) $p < 0.05$ compared to 37°C, plus rocking and CO_2, and 37°C, no rocking plus CO_2.

3. Several parameters of blood incubation were evaluated. The blood was handled in four different conditions, all with LPS stimulation. The first placed the blood on a rotating platform at 37°C in a 5% CO_2 incubator, the second placed the tubes vertically in the same incubator but without constant rotation, the third placed the tubes vertically in a 37°C bacteriological incubator without rotation or CO_2, and for the fourth condition the tubes were merely placed on the bench top, i.e., room temperature. Biologically active TNF or immunoreactive IL-8 was measured 24 h later in the plasma. For TNF production, only temperature was critical for the optimal production (**Fig. 2A**). However, the correct temperature and mixing was needed to ensure optimal production of IL-8 (**Fig. 2B**). CO_2 is not needed, which means that in remote locations a simple incubator is sufficient. The need for mixing to obtain optimal IL-8 production undoubtedly relates to the kinetics of cytokine production. TNF, similar to IL-6, is mostly produced during the first 8–12 h, whereas IL-8 production continues to increase throughout the length of the stimulation.
4. The plasma may be used for a variety of purposes such as measuring the cytokines. Either cytokine bioassays or immunoassays may be performed. As this plasma is already anticoagulated it may be used directly in either assay.

Acknowledgments

This work is supported in part by National Institutes of Health grants 44918 and GM 50401. J.S.F.'s whole-blood studies are in part supported by The Wellcome Trust.

References

1. Remick, D. G. and Friedland, J. S., eds. (1997) *Cytokines in Health and Disease*, 2 ed. Marcel Dekker, New York, NY.
2. Remick, D. G. (1995) Applied molecular biology of sepsis. *J. Crit. Care* **10,** 198–212.
3. O'Shea, J. J., Notarangelo, L. D., Johnston, J. A., and Candotti, F. (1997) Advances in the understanding of cytokine signal transduction: the role of Jaks and STATs in immunoregulation and the pathogenesis of immunodeficiency. *J. Clin. Immunol.* **17,** 431–447.
4. Suffredini, A. F., Reda, D., Banks, S. M., Tropea, M., Agosti, J. M., and Miller, R. (1995) Effects of recombinant dimeric TNF receptor on human inflammatory responses following intravenous endotoxin administration. *J. Immunol.* **155,** 5038–5045.
5. Martich, G. D., Danner, R. L., Ceska, M., and Suffredini, A. F. (1991) Detection of interleukin 8 and tumor necrosis factor in normal humans after intravenous endotoxin: the effect of antiinflammatory agents. *J. Exp. Med.* **173,** 1021–1024.
6. Kasahara, K., Strieter, R. M., Chensue, S. W., Standiford, T. J., and Kunkel, S. L. (1991) Mononuclear cell adherence induces neutrophil chemotactic factor/interleukin-8 gene expression. *J. Leukocyte Biol.* **50,** 287–295.

7. Doherty, J. F., Golden, M. H., Remick, D. G., and Griffin, G. E. (1994) Production of interleukin-6 and tumour necrosis factor-alpha in vitro is reduced in whole blood of severely malnourished children. *Clin. Sci. (Colch).* **86,** 347–351.
8. Kremer, J. P., Jarrar, D., Steckholzer, U., and Ertel, W. (1996) Interleukin-1, -6 and tumor necrosis factor-alpha release is down-regulated in whole blood from septic patients. *Acta Haematologica* **95,** 268–273.
9. van der Poll, T., Coyle, S. M., Moldawer, L. L., and Lowry, S. F. (1996) Changes in endotoxin-induced cytokine production by whole blood after in vivo exposure of normal humans to endotoxin. *J. Infec. Dis.* **174,** 1356–1360.
10. Mitov, I. G., Kropec, A., Benzing, A., Just, H., Garotta, G., Galanos, C., and Freudenberg, M. (1997) Differential cytokine production in stimulated blood cultures from intensive care patients with bacterial infections. *Infection* **25,** 206–212.
11. Friedland, J. S., Hartley, J. C., Hartley, C. G., Shattock, R. J., and Griffin, G. E. (1996) Cytokine secretion in vivo and ex vivo following chemotherapy of Mycobacterium tuberculosis infection. *Trans. Royal So. Trop. Med. Hygiene* **90,** 199–203.
12. Majetschak, M., Flach, R., Heukamp, T., Jennissen, V., Obertacke, U., Neudeck, F., Schmit-Neuerburg, K. P., and Schade, F. U. (1997) Regulation of whole blood tumor necrosis factor production upon endotoxin stimulation after severe blunt trauma. *J. Trauma* **43,** 880–887.
13. Dahlke, E., Schlag, R., Langenmayer, I., Frankenberger, M., Kafferlein, E., Subkowski, T., Emmerich, B., and Ziegler-Heitbrock, H. W. (1995) Decreased production of TNF and IL-6 in whole blood of CLL patients. *Am. J. Hematol.* **49,** 76–82.
14. Fischer, J. R., Schindel, M., Stein, N., Lahm, H., Gallati, H., Krammer, P. H., and Drings, P. (1995) Selective suppression of cytokine secretion in patients with small-cell lung cancer. *Ann. Oncol.* **6,** 921–926.
15. Elsasser-Beile, U., von Kleist, S., Stahle, W., Schurhammer-Fuhrmann, C., Monting, J. S., and Gallati, H. (1993) Cytokine levels in whole blood cell cultures as parameters of the cellular immunologic activity in patients with malignant melanoma and basal cell carcinoma. *Cancer* **71,** 231–236.
16. Elsasser-Beile, U., von Kleist, S., Fischer, R., and Monting, J. S. (1992) Impaired cytokine production in whole blood cell cultures from patients with colorectal carcinomas as compared to benign colorectal tumors and controls. *J. Clin. Lab. Anal.* **6,** 311–314.
17. Zangerle, P. F., De Groote, D., Lopez, M., Meuleman, R. J., Vrindts, Y., Fauchet, F., Dehart, I., Jadoul, M., Radoux, D., and Franchimont, P. (1992) Direct stimulation of cytokines (IL-1 beta, TNF-alpha, IL-6, IL-2, IFN-gamma and GM-CSF) in whole blood: II. Application to rheumatoid arthritis and osteoarthritis. *Cytokine* **4,** 568–575.
18. Chofflon, M., Juillard, C., Gauthier, G., and Grau, G. E. (1991) Correlation between in vitro cytokine production and clinical evolution of multiple sclerosis patients. *Schweiz. Arch. Neurol. Psychiatr.* **142,** 107–112.

19. Chofflon, M., Roth, S., Juillard, C., Paunier, A. M., Juillard, P., Degroote, D., and Grau, G. E. (1997) Tumor necrosis factor production capacity as a potentially useful parameter to monitor disease activity in multiple sclerosis. *Eur. Cytokine Network* **8,** 253–257.
20. Hautecoeur, P., Forzy, G., Gallois, P., Demirbilek, V., and Feugas, O. (1997) Variations of IL2, IL6, TNF alpha plasmatic levels in relapsing remitting multiple sclerosis. *Acta Neurologica Belgica* **97,** 240–243.
21. Evans, C. A., Jellis, J., Hughes, S. P., Remick, D. G., and Friedland, J. S. (1998) Tumor necrosis factor-alpha, interleukin-6, and interleukin-8 secretion and the acute-phase response in patients with bacterial and tuberculous osteomyelitis. *J. Infect. Dis.* **177,** 1582–1587.
22. van der Poll, T., Jansen, J., Endert, E., Sauerwein, H. P., and van Deventer, S. J. (1994) Noradrenaline inhibits lipopolysaccharide-induced tumor necrosis factor and interleukin 6 production in human whole blood. *Infec. Immun.* **62,** 2046–2050.
23. van der poll, T. and Lowry, S. F. (1997) Lipopolysaccharide-induced interleukin 8 production by human whole blood is enhanced by epinephrine and inhibited by hydrocortisone. *Infect. Immun.* **65,** 2378–2381.
24. Hartman, D. A., Ochalski, S. J., and Carlson, R. P. (1995) The effects of anti-inflammatory and antiallergic drugs on cytokine release after stimulation of human whole blood by lipopolysaccharide and zymosan A. *Inflamm. Res.* **44,** 269–274.
25. Chan, B., Kalabalikis, P., Klein, N., Heyderman, R., and Levin, M. (1996) Assessment of the effect of candidate anti-inflammatory treatments on the interaction between meningococci and inflammatory cells in vitro in a whole blood model. *Biotherapy* **9,** 221–228.
26. Matsumoto, N., Aze, Y., Akimoto, A., and Tujita, T. (1998) ONO-4007, an anti-tumor lipid A analog, induces tumor necrosis factor-alpha production by human monocytes only under primed state: different effects of ONO-4007 and lipopolysaccharide on cytokine production. *J. Pharmacol. Exp. Ther.* **284,** 189–195.
27. Redl, H., Bahrami, S., Leichtfried, G., and Schlag, G. (1992) Solving one of the problems of in vitro cytokine production and endotoxin measurement in blood [letter]. *Clin. Chim. Acta.* **205,** 149–150.
28. DeForge, L. E., Tracey, D. E., Kenney, J. S., and Remick, D. G. (1992) Interleukin-1 receptor antagonist protein inhibits interleukin-8 expression in lipopolysaccharide-stimulated human whole blood. *Am. J. Pathol.* **140,** 1045–1054.
29. DeForge, L. E. and Remick, D. G. (1991) Kinetics of TNF, IL-6, and IL-8 gene expression in LPS- stimulated human whole blood. *Biochem. Biophys. Res. Comm.* **174,** 18–24.
30. DeForge, L. E., Kenney, J. S., Jones, M. L., Warren, J. S., and Remick, D. G. (1992) Biphasic production of IL-8 in lipopolysaccharide (LPS)-stimulated human whole blood. Separation of LPS- and cytokine-stimulated components using anti-tumor necrosis factor and anti- IL-1 antibodies. *J. Immunol.* **148,** 2133–2141.

31. DeForge, L. E., Fantone, J. C., Kenney, J. S., and Remick, D. G. (1992) Oxygen radical scavengers selectively inhibit interleukin 8 production in human whole blood. *J. Clin. Invest.* **90,** 2123–2129.
32. DeForge, L. E., Preston, A. M., Takeuchi, E., Kenney, J., Boxer, L. A., and Remick, D. G. (1993) Regulation of interleukin 8 gene expression by oxidant stress. *J. Biol. Chem.* **268,** 12–13.
33. Cale, D. R. and Remick, D. G. (1998) Low molecular weight heparin is associated with greater cytokine production in stimulated whole blood model. *Shock* **10,** 192–197.

4

Nitric Oxide and Other Reactive Nitrogen Intermediates

11

Nitric Oxide Synthase Inhibitors

Dilani K. Siriwardena, Hajime Tagori,
and Christoph Thiemermann

1. Introduction

In 1990, several studies reported that an enhanced formation of endogenous nitric oxide (NO) contributes to the hypotension caused by endotoxin and tumor-necrosis factor-α (TNFα) *(1–3)*. In addition, it became apparent that this overproduction of NO also plays an important role in the pathophysiology of the vascular hyporesponsiveness to vasoconstrictor agents (also termed vasoplegia) *(4,5)*. Since then, many studies have reported on the effects and side effects of NO synthase (NOS) inhibitors in animal models of shock. Prior to discussing the pharmacology of various classes of NOS inhibitors, this chapter will briefly introduce the physiological role(s) of NO as well as the various (beneficial as well as detrimental) roles of NO in animal models of septic shock.

NO is generated from L-arginine by a family of enzymes collectively called NO synthases (NOSs), which contain an oxygenase domain (containing the catalytic center) and a reductase domain. The synthesis of NO from L-arginine and molecular oxygen involves the generation of N^G-hydroxy-L-arginine and water (first step) and subsequently the oxidation of N^G-hydroxy-L-arginine in the presence of molecular oxygen to form NO, L-citrulline, and water. When generated, NO diffuses to adjacent cells where it activates soluble guanylate cyclase, resulting in the formation of cGMP, which in turn mediates many (but not all) of the effects of NO. NO is generated by many mammalian cells by at least three different isoforms of NOS. The NOS in endothelial cells (eNOS or NOS III) and neuronal cells (nNOS or NOS I) are expressed constitutively, and both enzymes require an increase in intracellular calcium for activation. Activation of macrophages and many other cells with pro-inflammatory cytokines

From: *Methods in Molecular Medicine, Vol. 36: Septic Shock*
Edited by: T. J. Evans © Humana Press Inc., Totowa, NJ

or endotoxin results in the expression of a distinct isoform of NOS (inducible NOS; iNOS or NOS II), the activity of which is functionally independent on changes in intracellular calcium *(7–11)*. Thus, it is not surprising that NO has many biological functions in the cardiovascular, nervous, and immune systems. For instance, activation of eNOS by sheer stress results in a continuous release of picomolar amounts of NO, which helps to regulate blood pressure and organ blood flow by causing vasodilatation and opposing the effects of circulating catecholamines. NO also reduces the adhesion of platelets and polymorphonuclear leukocytes (PMNs) to the endothelium *(8)*. The latter effect of NO is, at least in part, caused by the prevention by NO of the expression of the adhesion molecules P-selectin and intercellular adhesion molecule (ICAM-1) on the surface of endothelial cells. In addition to preventing the adhesion of platelets to endothelial cells, NO also attenuates the activation of platelets directly. These effects of NO are associated with and/or caused by prevention of the expression of P-selectin (on platelets), secretion of platelet granules, intracellular calcium flux, as well as binding of glycoprotein IIb/IIIa to fibrinogen *(12)*.

The overproduction of NO in animal models of circulatory shock is caused by an early activation of eNOS (which is transient) and the delayed induction of iNOS activity (resulting in the formation of nanomolar amounts of NO) in macrophages (host defence), vascular smooth muscle (hypotension, vascular hyporeactivity, maldistribution of blood flow), and parenchymal cells *(13)*. The finding that inhibitors of NOS activity attenuate the hypotension and vasoplegia caused by endotoxin in animals, together with the discovery that mice in which the iNOS gene has been inactivated by gene targeting (iNOS-knockout mice) exhibit only a minor fall in blood pressure when challenged with endotoxin *(14,15)* support the hypothesis that an overproduction of NO by iNOS contributes to the circulatory failure in septic shock. It is less clear whether increased formation of NO also contributes to the organ injury and dysfunction caused by endotoxin. The formation of NO by eNOS (and potentially also by iNOS) also exerts beneficial effects in shock including vasodilatation, prevention of platelet and leukocyte adhesion, maintenance of microcirculatory blood flow and augmentation of host defence. Thus, it is not surprising that basic and clinical scientists have advocated the use of contrasting therapeutic approaches including inhibition of NOS activity, enhancement of the availability of NO (NO-donors, NO-inhalation) or a combination of both approaches. The following paragraphs highlight some of the effects and side effects of inhibitors of NOS activity in animal models of septic shock.

2. Materials

In all of our studies aimed at elucidating the effects and side effects of various inhibitors of NOS activity on hemodynamics and multiple organ dysfunction caused by endotoxin in the rat, we have used bacterial lipopolysaccharide from *Escherichia coli* of the serotype 0.127:B8. This endotoxin can be obtained form Sigma (Poole, Dorset, UK or St. Louis, MO). The NOS inhibitors L-N^G-nitroarginine methyl ester (L-NAME, cat. no. 105/003), N^G-methyl-L-arginine (L-NMMA, cat. no. 106/001), aminoguanidine (cat. no. 420/006), L-N^{ω}-(1-Iminoethyl)-L-lysine (L-NIL, cat. no. 270/010), and 1400W (cat. no. 270/073) can be obtained from Alexix Corporation (Nottingham, UK). S-methyl-isothiourea (cat. no. M3127) or aminoethyl-isothiourea (cat. no. A5879) can be obtained from Sigma.

3. Methods

As the actual addition of the drugs to be discussed is technically extremely easy, the main focus of the methods section will differ from the layout elsewhere in the book, and will focus on a discussion of the various factors that are important in considering a given drug.

When choosing and using inhibitors of NOS, one needs to have a good understanding of the pharmacology of the currently available NOS inhibitors. For instance, NOS inhibitors differ in their selectivity as inhibitors of the various NOS isoforms, in their mechanism of NOS inhibition, their bioavailability, metabolism, cellular uptake, and tissue distribution (in vivo). In addition to understanding the pharmacology, one needs to have good understanding of the pathophysiology of the disease in question to design a rational pharmacotherapy. For instance, one needs to determine whether one wishes to block, e.g., the generation of endogenous NO from all isoforms or, alternatively, whether one wishes to reduce the generation of NO by one particular NOS isoform. All of these questions have to be addressed prior to selecting an NOS inhibitor for a particular study. There is evidence that an enhanced formation of NO by iNOS may contribute to the pathophysiology of septic shock and other diseases (e.g., multiple sclerosis, inflammation, and so on) and also that the inhibition of eNOS activity may lead to substantial adverse effects. The following sections will review the pharmacology of various, chemically distinct NOS inhibitors and point out advantages and disadvantages of specific agents in animal models of septic or endotoxic shock.

3.1. L-NAME

L-NAME is a potent inhibitor of NOS activity of all isoforms. When administered in lower concentrations/doses, L-NAME functions as a relatively selective inhibitor of eNOS activity. Many studies that have employed L-NAME in vivo to inhibit iNOS activity have used doses of this NOS inhibitor that have caused a very substantial inhibition of eNOS activity. Thus, it is not surprising that many of these studies that have been reviewed recently ***(16)*** have demonstrated that L-NAME may cause a substantial degree of adverse effects when given in animal models of shock. These include excessive vasoconstriction (particularly in the pulmonary, renal, splanchnic, and myocardial vascular beds) and an increase in the incidence of both microvascular thrombosis and neutrophil adhesion to the endothelium. Thus, L-NAME reduces oxygen delivery and exacerbates organ injury in (many, but not all) animal models of endotoxic or septic shock. These results are not necessarily solely owing to the use of very large amounts of L-NAME, but rather a reflection of the fact that L-NAME is a more selective inhibitor of eNOS than iNOS activity. In rats with endotoxemia, even infusion of very low doses of L-NAME (e.g., 0.03–0.3 mg/kg/h) result in a dose-related increase in blood pressure (because of inhibition of eNOS activity) without reducing the rise in the plasma levels of nitrite/nitrate (an indicator of iNOS activity) or the organ injury caused by endotoxin *(17)*.

The hypothesis that the basal release of NO by eNOS has an important role in the regulation of regional blood flow and adhesion of blood-borne cells to the endothelium (beneficial effects of NO), whereas the excessive generation of NO by iNOS "in the wrong place at the wrong time" contributes to some aspects of the pathophysiology of shock (harmful effects of NO), has stimulated the search for selective inhibitors of iNOS activity. The following paragraphs highlight some aspects of the beneficial and adverse effects of relative selective inhibitors of iNOS activity in animal models of endotoxemia and sepsis (*see* **Table 1** for overview of different NOS inhibitors and comparison of their pharmacology, selectivity and potency).

3.2. L-NMMA

The N-substituted L-arginine analogue, L-NMMA, was the first agent reported to inhibit NOS activity. Although L-NMMA inhibits all isoforms of NOS to a variable degree, it is a more potent inhibitor of iNOS than eNOS activity in cultured cells *(18)*. L-NMMA is a competitive inhibitor of the binding of L-arginine to NOS and, hence, excess of L-arginine reverses the inhibition of NOS activity by L-NMMA. It should, however, be noted that prolonged exposure of iNOS to L-NMMA results in a time-dependent irreversible inactivation of enzyme activity, which is preceded by an nicotinamide adenine dinucleotide phosphate (NADPH)-dependent hydroxylation of the inhibitor to N-hydroxy-N-methyl-L-arginine *(19)*. As L-NMMA is only a moderately selective inhibitor of iNOS activity, it is not entirely surprising that effects of L-NMMA in models of shock vary from "very beneficial" (inhibition of iNOS activity) to "moderately beneficial with some adverse effects" (inhibition of eNOS activity masking the beneficial effects of iNOS inhibition) to "detrimental" (marked inhibition of eNOS activity). Clearly, the observed result is very

Table 1
Pharmacology of Some of the Commonly Used Inhibitors of Nitric Oxide Synthase

Inhibitor	Comments
N^G-methyl-L-arginine	Prototypical NOS inhibitor, which does not discriminate between the different NOS isoforms (e.g., nonselective NOS inhibitor)
N^G-nitro-L-arginine	Potent (and relatively selective inhibitor) of the constitutive NOS isoforms (eNOS and bNOS)
N^6-(1-iminoethyl)-L-lysine	Potent and selective inhibitor of iNOS in vitro and in vivo
S-methyl-isothiourea	Potent inhibitor of all NOS isoforms which exhibits some selectivity for iNOS
S-ethyl-isothiourea	Potent, but nonselective inhibitor of all isoforms of NOS
S-aminoethyl-isothiourea	Selective inhibitor of iNOS activity, prodrug for the formation of mercaptoethylguanidine (selective iNOS inhibitor, but also radical scavenger)
Aminoguanidine	Selective inhibitor of iNOS activity, but not very potent, which exerts a wide variety of nonspecific effects (unrelated to NOS inhibition)
1400W	Potent (in human), iNOS selective inhibitor of NOS activity, irreversible inhibition of iNOS activity has been described

Please note that the potency and IC_{50} of the above NOS inhibitors vary between species and tissues used.

dependant on the dose of L-NMMA as well as the model of shock (e.g., species, degree of iNOS induction, and so on). When given after the onset of hypotension following the injection of endotoxin, infusions of relatively low doses of L-NMMA (3–10 mg/kg/h) have been demonstrated to exert beneficial hemodynamic effects in rodent, sheep and baboon models of endotoxemia and sepsis. It should, however, be noted that the administration of larger amounts of L-NMMA (or presumably also the administration of smaller amounts of L-NMMA for several days) may be associated with side effects secondary to the inhibition of eNOS activity. For instance, there is evidence that L-NMMA enhances the degree of microthrombosis in the renal vasculature of rats with endotoxin shock *(20)*.

In contrast to rodents, sheep are very sensitive to small doses of endotoxin in a manner similar to humans. Indeed, infusion of either endotoxin or bacteria into sheep leads (within 24 h) to a hyperdynamic circulation with a fall in peripheral vascular resistance, an increase in cardiac output, increases in organ blood flow associated with a reduction in oxygen extraction. In this model, prolonged periods of endotoxemia or bacteremia (*Pseudomonas aeruginosa*) are also associated with increases in total renal blood flow and the development in precapillary arterio-venous shunts resulting in regional maldistribution of renal blood flow, fall in glomerular filtration pressure, and ultimately glomerular filtration rate. Interestingly, administration (at 24 h after the onset of endotoxemia) of L-NMMA increased urine output and reversed the impairment in creatinine clearance caused by infusion of bacteria, without causing a significant fall (below baseline) of renal blood flow. In addition to these beneficial effects on renal blood flow and function, NOS inhibition also resulted in an increase in oxygen extraction, a fall in organ blood flow from elevated to normal levels (in brain, heart, jejunum, ileum), an increase in peripheral vascular resistance, but no significant increase in lactate, indicating a normalization of hemodynamic parameters in the absence of excessive vasoconstriction *(21,22)*. In primates, administration of live *E. coli* bacteria (2×10^9 colony forming units) resulted (after 4 h) in a significant increase in the serum levels of biopterin, neopterin, and nitrate, suggesting that administration of live bacteria results in the induction of GTP cyclohydrolase I and iNOS within 4 h *(23)*. In this model, infusion of L-NMMA (5 mg/kg/h) attenuated the rise in the serum levels of nitrate and creatinine, the hypotension and fall in peripheral vascular resistance, and the substantial 7-day mortality caused by severe sepsis in this species (D. Rees and H. Redl, pers. comm.). These findings clearly document that the circulatory failure caused by septic shock in baboons is mediated largely by an enhanced formation of NO by NOS and that inhibition of NOS activity with L-NMMA (or other more selective inhibitors of iNOS activity) improves outcome in this model.

Taken together, L-NMMA is (in higher concentrations) a nonselective inhibitor of NOS activity and, hence, may cause adverse effects due to excessive inhibition of eNOS activity. As L-NMMA is a more potent inhibitor of iNOS than eNOS activity, it is possible to select doses of L-NMMA which inhibit iNOS activity, but only cause a minor inhibition of eNOS activity. It is not clear whether it is possible to use L-NMMA as a tool to selectively inhibit iNOS activity in chronic (days rather than hours) models of shock.

3.3. Aminoguanidine (and Other Guanidines)

Aminoguanidine was the first relatively selective inhibitor of iNOS activity discovered *(24)*. Although aminoguanidine is a more potent inhibitor of iNOS than eNOS activity in vitro and in vivo *(25–28)*, aminoguanidine is not a very potent inhibitor of iNOS activity (IC_{50} approx 100–150 μM). The inhibition of

NOS by aminoguanidine becomes greater with increasing incubation time indicating that aminoguanidine is a mechanism-based inhibitor *(29)*. In addition to aminoguanidine, other guanidines including (in the rank order of their potency as inhibitors of iNOS activity in murine macrophages and smooth muscle cells) 1-amino-2-hydroxy-guanidine (IC_{50} approx 70–100 μM), 1-amino-2-methyl-guanidine, 1-amino-1-methyl-guanidine, and 1-amino-1,2-dimethylguanidine also inhibit iNOS activity *(30)*. Of these, 1-amino-2-hydroxy-guanidine is more potent, more selective, and better soluble than aminoguanidine itself and, hence, may be more suitable than the respective parent compound *(30)*.

The findings that aminoguanidine attenuates the vascular hyporeactivity to phenylephrine in the aorta and main pulmonary artery of rats treated with endotoxin *(26,27)*, without affecting vascular tone or endothelium-dependent relaxations, provided support for the concept that aminoguanidine may exert beneficial effects in shock. Indeed, aminoguanidine attenuates the delayed hypotension in rats *(28)* and rabbits *(31)* with endotoxin shock and improves survival in mice challenged with endotoxin *(28)*. We have evaluated recently the effects of aminoguanidine and 1-hydroxy-2-guanidine on hemodynamics and organ dysfunction/injury in rats with sever endotoxemia. Endotoxemia for 6 h resulted in circulatory failure comprising of hypotension, tachycardia, and a vascular hyporeactivity of the vasculature to the pressor responses elicited by norepinephrine. This was associated with liver dysfunction and hepatocellular injury, acute renal failure, and induction of iNOS activity. Therapeutic administration (e.g., continuous infusion starting at 2 h after injection of endotoxin) of aminoguanidine *(28)* or 1-hydroxy-2-amino-guanidine *(30)* attenuated the liver dysfunction and hepatocellular injury which developed between 4 and 6 h after the injection of endotoxin. In the rat, endotoxin also causes a decrease in total cytochrome P450 content (liver) and a reduction of the metabolism of ethylmorphin and metazolam; and both of these effects are diminished by aminoguanidine *(32)*. In contrast, aminoguanidine did neither attenuate the impairment in gluconeogenesis *(33)*, nor the cardiac depresssion caused by endotoxin in the rat *(34)*. Aminoguanidine also decreased the degree of bacterial translocation presumably by preventing the injury to the gut mucosal barrier *(35)*, the disruption of the blood–brain barrier *(36)* as well as the increase in pulmonary transvascular flux *(37)* caused by endotoxin in the rat.

Unfortunately, the interpretation of the mechanism(s) by which aminoguanidine exerts its beneficial effects in shock (and other diseases) is extremely difficult, as aminoguanidine is not a very specific inhibitor of iNOS activity. Indeed, aminoguanidine has many other pharmacological properties including inhibition of histidine decarboxylase *(38)*, polyamine catabolism

(39), the formation of advanced glycosylation end products, and of catalase activity (as well as other copper- or iron-containing enzymes) *(41)*. Moreover, aminoguanidine inhibits the oxidative modification of low-density lipoprotein and the subsequent increase in uptake by macrophage scavenger receptor *(41)*. Interestingly, aminoguanidine also prevents the expression of iNOS protein by a hitherto unknown mechanism *(42)*.

Thus, aminoguanidine has to be regarded as an agent that is a relatively selective, but not very potent inhibitor of iNOS activity; reduces the formation of NO by two distinct mechanisms, namely prevention of the expression of iNOS protein and inhibition of iNOS activity; and exerts many other effects that appear to be unrelated to the inhibition of iNOS activity (nonspecific effects).

3.4. S-Substituted Isothioureas (e.g., Aminoethyl-Isothiourea)

S-substituted isothioureas (ITUs) are non-amino acid analogs of L-arginine and also potent inhibitors of iNOS activity with variable isoform selectivity *(43–45)*. The most potent isothioureas are those with only short alkyl chains on the sulphur atom and no substituents on the nitrogen atoms. For instance, S-ethyl-ITU is a potent competitive inhibitor of human iNOS (K_i value: 17 n*M*), eNOS (K_i value: 36 n*M*), and bNOS (K_i value: 29 n*M*) as well as murine iNOS (K_i value: 5 n*M*). S-ethyl-ITU binds near the heme of iNOS, the sulphur atom plays a critical role in binding, and the small S-ethyl substituent may well fit into a small hydrophobic pocket on the enzyme *(43)*. In contrast to S-ethyl-ITU, aminoethyl-ITU and S-methyl-ITU are more selective inhibitors of iNOS than of eNOS activity in vitro and in vivo *(45)*. Interestingly, aminoethyl-ITU is metabolized to mercaptoethyl-guanidine (and possibly the isothiourea, 2-aminothiazoline), which may represent the active principle of aminoethyl-ITU *(46)*. Indeed, mercaptoalkyl-guanidines, in particular mercaptoethyl- and mercaptopropyl-guanidines, are the most potent (guanidino) inhibitors of iNOS activity *(9)*. Although S-substituted isothioureas also inhibit nNOS activity, they do not cross the blood–brain barrier and, hence, may well be useful tools to elucidate the role of an enhanced formation of NO from iNOS in disease states including circulatory shock and inflammation.

S-methyl-ITU reverses the hypotension and the vascular hyporeactivity to norepinephrine caused by endotoxin in a dose-dependent fashion (0.01–3 mg/kg iv). In addition, therapeutic administration of S-methyl-ITU (5 mg/kg at 2 h after injection of endotoxin) also reduced the rises in the plasma levels of alanine aminotransferase (ALT), aspartate aminotransferase (AST), bilirubin, and creatinine at attenuated the hypocalcemia caused by 6 h of endotoxemia *(44)*. Moreover, S-methyl-ITU reduced the 24-h mortality caused by severe endotoxemia in mice *(44)*. Although S-methyl-ITU selectively inhibited iNOS activity in vitro, it did not affect the activity of xanthine oxidase, diaphorase, catalase, superoxide dismutase, lactate dehydrogenase, or monoamine oxidase *(44)*. In blood vessels obtained from rats with endotoxemia, S-methyl-ITU also attenuated the vasodilator effect and the rise in vascular cGMP caused by L-arginine *(47)*. Moreover, S-methyl-ITU also reduced the vascular leak in the pulmonary vasculature of rats with endotoxic shock *(37)*.

The finding that aminoethyl-ITU also reduced the delayed hypotension in rats with endotoxemia (inhibition of iNOS activity), without increasing blood pressure in rats that had not received endotoxin (inhibition of eNOS activity) supports the view that aminoethyl-ITU is a selective inhibitor of iNOS activity in vivo. Therapeutic administration of aminoethyl-ITU (1 mg/kg/h commencing 2 h after injection of endotoxin) resulted in beneficial hemodynamic effects and attenuated the degree of liver injury/dysfunction caused by endotoxin in the rat *(48)*. In pigs with endotoxemia, injection of aminoethyl-ITU (10 mg/kg iv at 3 h after endotoxin) restored hepatic arterial blood flow (from reduced to normal levels) and increased hepatic oxygen consumption, without affecting cardiac output *(49)*. S-ethyl-ITU has been reported to attenuate the plasma extravasation caused by endotoxin in mice *(50)* and to restore blood pressure in rabbits with endotoxemia *(31)*. It is, however, unclear whether these effects are caused by inhibition of iNOS activity as S-ethyl-ITU is also a potent inhibitor of eNOS activity *(44)*. Similarly, low doses of S-isopropyl-ITU (a potent inhibitor of eNOS activity) exert beneficial hemodynamic effects in animal models of hemorrhagic shock. For instance, in a porcine model of hemorrhagic shock, S-isopropyl-ITU (0.3 mg/kg bolus plus 1 mg/kg/h infusion) increased blood pressure and systemic vascular resistance, but did not alter cardiac index, renal blood flow, arterial or portal oxygen content or splanchnic oxygen consumption or extraction. Moreover, the increase in blood pressure afforded by S-isopropyl-ITU in rats subjected to severe hemorrhage was associated with a prolongation of survival *(51)*.

Having stressed that some of the beneficial effects of aminoguanidine in shock may not be caused by its ability to inhibit iNOS activity (e.g., nonspecific effects), it should be noted that the isothioureas mentioned above are also likely to elicit effects that are unrelated to their ability to inhibit NOS activity. For instance, aminoethyl-ITU is a radical scavenger and exert beneficial effects in models of disease/pathology known to be mediated by oxygen-derived free radicals *(50,51)*. Moreover, aminoethyl-ITU also prevents the expression of iNOS protein caused by endotoxin in cultured macrophages and in the rat in vivo *(42)*. Although similar, nonspecific effects have not been reported for S-methyl-ITU, the latter isothiourea is a more potent inhibitor of eNOS activity than aminoethyl-ITU and will therefore (at higher doses/concentrations) also inhibit eNOS activity.

3.5. 1400W (and Other Amidines)

S-substituted ITUs and guanidines contain the amidine function (-CH[=NH]NH_2), a feature which they have in common with O-substituted isoureas and amidines themselves. Indeed, amidines including 2-iminopiperidine, butyramine, 2-aminopyridine, propioamidine, and acetamidine inhibit NOS activity. Interestingly, both 2-iminopiperidine and butyramidine were more potent inhibitors of iNOS activity than L-NMMA in murine macrophages *(52)*. Recently, an analog of acetamidine termed 1400W (N-[3-(aminomethyl)benzyl]acetamidine) has been reported to be a slow, tight-binding inhibitor of human iNOS. The inhibition by 1400W of the activity of human iNOS was extremely potent (K_d value approx 7 n*M*), dependent on the cofactor NADPH, and either irreversible or extremely slowly reversible. Most notably, 1400W was approx 5000-fold more potent as an inhibitor of iNOS activity than of eNOS activity. Moreover, the inhibition by this agent of the activity of eNOS or nNOS activity was reversible by L-arginine, whereas iNOS inhibition was not. In a rat model of vascular injury caused by endotoxin, 1400W was 50-fold more potent as an inhibitor of iNOS than eNOS activity *(53)*. Thus, 1400W appears to be the most potent and selective inhibitor of iNOS activity known to date and, hence, will be an ideal tool to elucidate the role of NO from iNOS in shock and other diseases associated with the induction of iNOS. We have discovered recently that 1400W (3-10 mg/kg/h) attenuates both iNOS activity as well as the delayed hypotension caused by endotoxin in the rat in a dose-related fashion. Most notably, infusion of 10 mg/kg/h of 1400W (starting 2 h after endotoxin) abolished both the increase in nitrite + nitrate as well as the delayed hypotension caused by LPS. These beneficial hemodynamic effects of 1400W did, however, not translate into a protection of organs such as the kidney, liver, and pancreas against the multiple-organ injury caused by endotoxin in the rat. These results suggest that an enhanced formation of NO from iNOS contributes to the circulatory failure, but not the organ dysfunction associated with endotoxic shock *(59)*.

3.6. L-N^6-(1-Iminoethyl)-L-Lysine (L-NIL)

In addition to 1400W, L-N^6-(l-iminoethyl)lysine (L-NIL) is a very selective inhibitor of iNOS activity *(54–58)* which in the rat inhibits the carageenan-induced paw edema *(55)* and plasma extravasation into the skin *(56)*, respectively. To date, no nonspecific effects (e.g., effects not secondary to inhibition of iNOS activity) of L-NIL have been reported. We have investigated recently the effects of L-NIL, on the hypotension, liver injury and dysfunction, renal dysfunction, pancreatic injury, and the rise in the serum levels of nitrite + nitrate in rats with endotoxemia. In animals without endotoxemia (sham-operated controls), infusion of L-NIL (3 mg/kg/h) did not result in any significant alterations in blood pressure, indicating that at the dose regime choosen L-NIL does not inhibit eNOS activity (which would result in an increase in blood pressure). Infusion of LPS (6 mg/kg iv over 15 min) resulted in a biphasic fall in MAP from 132 ± 3 mmHg (baseline, prior to infusion of LPS) to 83 ± 4 mmHg at 360 min, which was abolished by treatment of LPS rats with bolus plus infusions of the selective iNOS inhibitor L-NIL (3 mg/kg/h, 2 h after LPS). The notion that L-NIL (in the doses administered in this study) is indeed a selective inhibitor of iNOS activity is supported by our findings that L-NIL attenuated the rise in the plasma levels of nitrite + nitrate caused by endotoxin, which is known to be caused by induction of iNOS activity and that 1400W did not cause an increase in blood pressure (which is caused by inhibition of eNOS activity) in rats that had not received endotoxin. This finding reinforces the hypothesis that the enhanced formation of NO by iNOS contributes largely to the delayed circulatory failure associated with endotoxic shock.

Endotoxemia for 6 h resulted in rises in the serum levels of the transaminases ALT and AST and γGT demonstrating the development of liver injury and dysfunction. Treatment of LPS rats with L-NIL did not attenuate the liver injury/dysfunction caused by LPS in the rat. When compared to rats that had received vehicle rather than LPS, endotoxemia for 6 h resulted in rises in the serum levels of urea and creatinine, demonstrating the development of renal dysfunction. Treatment of LPS rats with L-NIL, however, did not attenuate the renal dysfunction caused by LPS in the rat. Similarly, L-NIL did also not affect the pancreatic dysfunction (measured as a rise in the serum levels of lipase) caused by endotoxin in the rat. Although L-NIL prevented the delayed hypotension caused by endotoxin, this iNOS inhibitor did not affect the development of the multiple organ injury and dysfunction caused by endotoxin. Possible explanations for this failure of L-NIL to attenuate the MODS include the following: Although blood pressure was restored by L-NIL, perfusion to vital organs may remain impaired, for example because of splanchnic vasoconstriction (in the case of the liver); enhanced formation of NO does not contribute significantly to endotoxin-induced organ injury; and L-NIL did not inhibit iNOS activity in the relevant target organs, although this is extremely unlikely, as the dose of L-NIL abolished the rise in nitrite/nitrate caused by LPS. Further studies are warranted to elucidate whether selective inhibition of iNOS activity attenuates the MODS caused by endotoxin (or bacteria) in large animal models of endotoxin shock and sepsis ***(59)***.

4. Notes

In the last years numerous agents have been reported to be more potent inhibitors of iNOS than eNOS activity, and some of these have been evaluated in animal models of endotoxemia. There is good evidence that inhibition of iNOS activity attenuates the circulatory failure in endotoxin (or septic) shock in many species. Although some inhibitors of iNOS activity reduce the organ injury/dysfunction associated with endotoxic or septic shock, it is still unclear whether these effects are related directly to the ability of these agents to inhibit iNOS activity, or secondary to improvements in hemodynamics (e.g., L-NMMA) or (at least in part) caused by nonspecific effects (e.g., aminoguanidine, aminoethyl-isothiourea). Studies using mice in which the iNOS gene has been inactivated by gene targeting to elucidate the role of iNOS in endotoxemia support the notion that NO from iNOS contributes to hypotension and host defense, but provide controversial results regarding the role of iNOS in organ injury and mortality *(14,15,60)*. We have discovered recently that the selective inhibition of iNOS activity with either 1400W or L-NIL attenuates the circulatory failure, but not the organ injury and dysfunction caused by endotoxin in the rat *(59)*. These findings support the view that selective inhibition of iNOS activity might a useful approach to the restoration of blood pressure in patients with shock. Our data are also consistent with the notion that—as in the case of iNOS knockout mice challenged with endotoxin—enhanced formation of NO by iNOS contributes primarily to the circulatory failure, but not to the MODS caused by endotoxin. Whether an improvement in hemodynamics produced by inhibition of iNOS activity will be sufficient to reduce vital organ injury and improve outcome in patients with septic shock warrants further investigation.

Acknowledgments

The author is a Senior Fellow of the British Heart Foundation (FS96/018).

References

1. Thiemermann, C. and Vane, J. R. (1990) Inhibition of nitric oxide synthesis reduces the hypotension induced by bacterial lipopolysaccharide in the rat. *Eur. J. Pharmacol.* **182,** 591–595.
2. Kilbourn, R. G., Juburan, A., Gross, S. S., Griffith, O. W., Levi, R., and Adams, J. (1990) Reversal of endotoxin-mediated shock by N^G-monomethyl-L-arginine, an inhibitor of nitric oxide synthesis. *Biochem. Biophys. Res. Commun.* **172,** 1132–1138.
3. Kilbourn, R. G., Gross, S. S., Jubran, A., Adams, J., Griffith, O. W., Levi, R., and Lodato, R. F. (1990) N^G-methyl-L-arginine inhibits tumour necrosis factor-

induced hypotension: implications for the involvement of nitric oxide. *Proc. Natl. Acad. Sci. USA* **87,** 3629–3632.

4. Julou-Schaeffer, G., Gray, G. A., Fleming, I., Schott, C., Parratt, J. R., and Stoclet, J. C. (1990) Loss of vascular responsiveness induced by endotoxin involves the L-arginine pathway. *Am. J. Physiol.* **259,** H1038–H1043.
5. Rees, D. D., Cellek, S., Palmer, R. M. J., and Moncada, S. (1990) Dexamethasone prevents the induction of nitric oxide synthase and the associated effects on the vascular tone: an insight into endotoxic shock. *Biochem. Biophys. Res. Commun.* **173,** 541–547.
6. Nathan, C. (1992) Nitric oxide as a secretory product of mammalian cells. *FASEB J.* **6,** 3051–3064.
7. Morris, S. M. and Billiar, T. R. (1994) New insights into the regulation of inducible nitric oxide synthase. *Am. J. Physiol.* **266,** E829–E839.
8. Moncada, S. and Higgs, A. (1993) The L-arginine-nitric oxide pathway. *N. Eng. J. Med.* **329,** 2202–2212.
9. Southan, G. J. and Szabo, C. (1996) Selective pharmacological inhibition of distinct nitric oxide synthase isoforms. *Biochem. Pharmacol.* **51,** 383–394.
10. Thiemermann, C. (1994) The role of L-arginine:nitric oxide pathway in circulatory shock. *Adv. Pharmacol.* **28,** 45–79.
11. Szabo, C. and Thiemermann, C. (1995) Regulation of the expression of the inducible isoform of nitric oxide synthase. *Adv. Pharmacol.* **34,** 113–154.
12. Loscalzo, J. and Welsch, G. (1995) Nitric oxide and its role in the cardiovascular system. *Proj. Cardiovasc. Dis.* **38,** 87–104.
13. Szabo, C. (1995) Alterations in nitric oxide production in various forms of circulatory shock. *Horizon* **1,** 2–32.
14. MacMicking, J. D., Nathan, C., Hom, G. et al. (1995) Altered responses to bacterial infection and endotoxic shock in mice lacking inducible nitric oxide synthase. *Cell* **81,** 641–650.
15. Wei, X., Charles, I. G., Smith, A. et al. (1995) Altered immune responses in mice lacking inducible nitric oxide synthase. *Nature* **375,** 408–411.
16. Kilbourn, R. G., Traber, D. L., and Szabo, C. (1997) Beneficial versus detrimental effects of nitric oxide synthase inhibitors in circulatory shock: lessons learned from experimental and clinical studies. *Shock* **7,** 235–246.
17. Wu, C. C., Ruetten, H., and Thiemermann, C. (1996) Comparison of the effects of aminoguanidine and N^G-nitro-L-arginine methyl ester on the multiple organ dysfunction caused by endotoxemia in the rat. *Eur. J. Pharmacol.* **300,** 99–104.
18. Gross, S. S., Stuehr, D. J., Aisaka, K., Jaffe, E. A., Levi. R., and Griffith, O. W. (1990) Macrophage and endothelial nitric oxide synthesis: cell-type selective inhibition by N^G-aminoarginine, N^G-nitroarginine and N^G-methyl-arginine. *Biochem. Biophys. Res. Commun.* **170,** 96–103.
19. Feldmann, P. L., Griffith, O. W., Honh, H., and Stuehr, D. (1993) Irreversible inactivation of macrophage and brain nitric oxide synthase by L-NG-methyl arginine requires NADPH-dependent hydroxylation. *J. Med. Chem.* **36,** 491–496.

20. Shultz, P. J. and Raij, L. (1992) Endogenously synthetized nitric oxide prevents endotoxin-induced glomerular thrombosis. *J. Clin. Invest.* **90,** 1718–1725.
21. Meyer, J., Traber, L. D., Nelson, S., Lentz, C. W., Nakazawa, H., Herndon, D. N., Noda, H., and Traber, D. L. (1992) Reversal of hyperdynamic response to continuous endotoxin administration by inhibition of NO synthesis. *J. Appl. Physiol.* **73,** 324–328.
22. Meyer, J., Lentz, C. W., Stothert, J. C., Traber, L. D., Herndon, D. N., and Traber, D. L. (1994) Effects of nitric oxide synthesis inhibition in hyperdynamic endotoxemia. *Crit. Care. Med.* **22,** 306–312.
23. Strohmeier, W., Werner, E. R., Redl, H., Wachter, H., and Schlag, G. (1995) Plasma nitrate and pteridine levels in experimental bacteremia in baboons. *Pteridines* **6,** 8–11.
24. Corbett, J. A., Tilton, R. G., Chang, K., Hasan, K. S., Ido, Y., Wang, J. L., Sweetland, M. A., Lancaster, J. R., Williamson, J. R., and McDaniel, M. L. (1992) Aminoguanidine, a novel inhibitor of nitric oxide formation, prevents diabetic vascular dysfunction. *Diabetes* **41,** 552–558.
25. Misko, T. P., Moore, W. M., Kasten, T. P., Nickols, D. A., Corbett, J. A., Tilton, R. G., McDaniel, M. L., Williamson, J. R., and Currie, M. G. (1993) Selective inhibition of the inducible nitric oxide synthase by aminoguanidine. *Eur. J. Pharmacol.* **233,** 119–125.
26. Griffith, M. J., Messent, M., MacAllister, R. J., and Evans, T. W. (1993) Aminoguanidine selectively inhibits inducible nitric oxide synthase. *Br. J. Pharmacol.* **110,** 963–968.
27. Joly, G. A., Ayres, M., Chelly, F., and Kilbourn, R. G. (1990) Effects of N^G-methyl-L-arginine, N^G-nitro-L-arginine and aminoguanidine on constitutive and inducible nitric oxide synthase in rat aorta. *Biochem. Biophys. Res. Commun.* **199,** 147–154.
28. Wu, C. C., Chen, S. J., Szabo, C., Thiemermann, C., and Vane, J. R. (1995) Aminoguanidine attenuates the delayed circulatory failure and improves survival in rodent models of endotoxic shock. *Br. J. Pharmacol.* **114,** 1666–1672.
29. Wolff, D. J. and Lubeskie, A. (1995) Aminoguanidine is an isoform-selective, mechanism-based inactivator of nitric oxide synthase. *Arch. Biochem. Biophys.* **316,** 290–301.
30. Ruetten, H., Southan, G. J., Abate, A., and Thiemermann, C. (1996) Attenuation of the multiple organ dysfunction caused by endotoxin by 1-amino-2-hydroxy-guanidine, a potent inhibitor of inducible nitric oxide synthase. *Br. J. Pharmacol.* **118,** 261–270.
31. Seo, H. G., Fujiwara, N., Kaneto, H., Asashi, M., Fujii, J., and Taniguchi, N. (1996) Effect of the nitric oxide synthase inhibitor, S-ethyl-isothiourea, on cultured cells and cardiovascular functions of normal and lipopolysaccharide-treated rabbits. *J. Biochem.* **119,** 553–558.
32. Muller, C. M., Scierka, A., Stiller, R. L., Kim, Y. M., Cook, D. R., Lancaster, J. R., Buffington, C. W., and Watkins, W. D. (1996) Nitric oxide mediates hepatic cytochrome P450 dysfunction induced by endotoxin. *Anesthesiology* **84,** 1435–1442.

33. Ou, J., Molina, L., Kim, Y. M., and Billiar, T. R. (1996) Excessive NO production does not account for the inhibition of hepatic gluconeogenesis in endotoxemia. *Am. J. Physiol.* **271,** 621–628.
34. Klabunde, R. E. and Coston, A. F. (1995) Nitric oxide synthase inhibition does not prevent cardiac depression in endotoxic shock. *Shock* **3,** 73–78.
35. Sorrells, D. L., Friend, C., Koltuksuk, U., Courcoulas, A., Boyle, P., Garrett, M., Watkins, S., Rowe, M. I., and Ford, H. R. (1996) Inhibition of nitric oxide with aminoguanidine reduces bacterial translocation after endotoxin challenge in vivo. *Arch. Surg.* **131,** 1155–1163.
36. Boje, K. M. (1996) Inhibition of nitric oxide synthase attenuates blood-brain barrier disruption during experimental meningitis. *Brain Res.* **720,** 75–83.
37. Arkovitz, M. S., Wispe, J. R., Garcia, V. F., and Szabo, C. (1996) Selective inhibition of the inducible isoform of nitric oxide synthase prevents pulmonary transvascular flux during acute endotoxemia. *J. Pediatr. Surg.* **31,** 1009–1015.
38. Bieganski, T., Kusche, J., Lorenz, W., Hesterberg, R., Stahlknecht, C. D., and Feussner, K. D. (1983) Distribution and properties of human diamine oxidase and its relevance for the histamine catabolism. *Biochem. Biophys. Acta* **756,** 196–203.
39. Seiler, N., Bolkenius, F. N., and Knodgen, B. (1985) The influence of catabolic reactions on polyamine excretion. *Biochem. J.* **225,** 219–226.
40. Ou, P. and Wolff, S. P. (1993) Aminoguanidine: a drug proposed for prophylaxis in diabetes inhibits catalase and generates hydrogen peroxide in vitro. *Biochem. Pharmacol.* **46,** 1139–1144.
41. Picard, S., Parthasarathy, S., Fruebis, J., and Witzum, J. L. (1992) Aminoguanidine inhibits oxidative modification of low density lipoprotein and the subsequent increase in uptake by macrophage scavenger receptors. *Proc. Natl. Acad. Sci. USA* **89,** 6876–6880.
42. Ruetten, H. and Thiemermann, C. (1986) Prevention of the expression of inducible nitric oxide synthase by aminoguanidine or aminoethylisothiourea in macrophages and in the rat. *Biochem. Biophys. Res. Commun.* **225,** 525–530.
43. Garvey, P. E., Oplinger, J. A., Tanoury, G. J., Sherman, P. A., Fowler, M., Marshall, S., Marmon, M. F., Paith, J. E., and Furfine, E. S. (1994) Potent and selective inhibition of human nitric oxide synthases. Inhibition by non-amino acid isothioureas. *J. Biol. Chem.* **269,** 26,669–26,676.
44. Szabo, C., Southan, G., and Thiemermann, C. (1994) Beneficial effects and improved survival in rodent models of septic shock with S-methyl-isothiourea sulfate, a novel, potent and selective inhibitor of inducible nitric oxide synthase. *Proc. Natl. Acad. Sci. USA* **91,** 12,472–12,476.
45. Southan, G., Szabo, C., and Thiemermann, C. (1995) Isothioureas: potent inhibitors of nitric oxide synthases with variable isoform selectivity. *Br. J. Pharmacol.* **114,** 510–516.
46. Southan, G. J., Zingarelli, B., O'Conner, M., Salzman, A. L., and Szabo, C. (1996) Spontaneous rearrangement of aminoalkylisothioureas into mercaptoalkylguanidines, a novel class of nitric oxide synthase inhibitors with selectivity towards the inducible isoform. *Br. J. Pharmacol.* **117,** 619–632.

47. Martinez, M. C., Muller, B., Stoclet, J. C., and Andriantsitohaina, R. (1996) Alteration by lipopolysaccharide of the relationship between intracellular calcium levels and contraction in rat mesenteric artery. *Br. J. Pharmacol.* **118,** 1218–1222.
48. Thiemermann, C., Ruetten, H., Wu, C. C., and Vane, J. R. (1995) The multiple organ dysfunction syndrome caused by endotoxin in the rat: Attenuation of liver dysfunction by inhibitors of nitric oxide synthase. *Br. J. Pharmacol.* **116,** 2845–2851.
49. Saetre, T., Gundersen, Y., Thiemermann, C., Lilleansen, P., and Aasen, A. O. (1998) Aminoethyl-isothiourea, a selective inhibitor of inducible nitric oxide synthase activity, improves liver circulation and oxygen metabolism in a porcine model of endotoxaemia. *Shock* **9,** 109–115.
50. Muraki, T., Fujii, E., Okada, M., Horikawa, H., Irie, K., and Ohba, K. (1996) Effect of S-ethyl-isothiourea, a putative inhibitor of inducible nitric oxide synthase, on mouse skin vascular permeability. *Jpn. J. Pharmacol.* **70,** 269–271.
51. Vromen, A., Szabo, C., Southan, G. J., and Salzman, A. L. (1996) Effects of S-isopropyl isothiourea, a potent inhibitor of nitric oxide synthase, in severe hemorrhagic shock. *J. Appl. Physiol.* **81,** 707–715.
52. Southan, G. J., Szabo, C., O'Conner, M. P., Salzman, A. C., and Thiemermann, C. (1995) Amidines are potent inhibitors of nitric oxide synthases: preferential inhibition of the inducible isoform. *Eur. J. Pharmacol.* **291,** 311–318.
53. Garvey, E. P., Oplinger, J. A., Furfine, E. S., Kiff, R. J., Laszlo, F., Whittle, B. J. R., and Knowles, R. G. (1997) 1400W is a slow, tight binding, and highly selective inhibitor of inducible nitric oxide synthase in vitro and in vivo. *J. Biol. Chem.* **272,** 4959–4963.
54. Moore, W. M., Webber, R. K., Jerome, G. M., Tjoeng, F. S., Misko, T. P., and Currie, M. G. (1994) L-N6-(l-iminoethyl)lysine: a selective inhibitor of inducible nitric oxide synthase. *J. Med. Chem.* 3886–3888.
55. Salvemini, D., Wang, Z. Q., Wyatts, P. S., Bourdon, D. M., Marino, M. H., Manning, P. T., and Currie, M. G. (1996) Nitric oxide: a key mediator in the early and late phase of carageenan-induced rat paw inflammation. *Br. J. Pharmacol.* **118,** 829–838.
56. Ridger, V. C., Pettipher, E. R., Bryant, C. E., and Brain, S. D. (1997) Effect of inducible nitric oxide synthase inhibitors aminoguanidine and L-N6-(l-iminoethyl)lysine on zymosan-induced plasma extravasation in rat skin. *J. Immunol.* **159,** 383–390.
57. Moralez-Ruiz, M., Jiminez, W., Ros, J., Sole, M., Leivas, A., Bosch-Marce, M., Rivera, F., Arroyo, V., and Rodes, J. (1997) Nitric oxide production in peritoneal macrophages of cirrhotic rats: a host response against bacterial peritonitis. *Gastroenterology* **112,** 2056–2064.
58. Faraci, W. S., Nagel, A. A., Verdies, K. A., Vincent, L. A., Xu, H., Nichols, L. E., Labasi, J. M., Salter, E. D., and Pettipher, E. R. (1996) 2-amino-4-pyrimidine is a potent inhibitor of inducible NO synthase activity in vitro and in vivo. *Br. J. Pharmacol.* **119,** 1101–1108.

59. Wray, G. M., Millar, C. G., Hinds, C. J., and Thiemermann, C. (1998) Selective inhibition of the activity of inducible nitric oxide synthase prevents the circulatory failure, but not the organ injury/dysfunction caused by endotoxin. *Shock* **9,** 329–335.
60. Laubach, V. E., Shesely, E. G., Smithies, O., and Sherman, P. A. (1995) Mice lacking inducible nitric oxide synthase are not resistant to lipopolysaccharide-induced death. *Proc. Natl. Acad. Sci. USA* **92,** 10,668–10,692.

12

In Situ Detection of Nitric Oxide

Tadeusz Malinski

1. Introduction

The measurement of the concentration of nitric oxide (NO) is a challenging problem because of the short half-life of NO ($t_{1/2}$ = 3–6 s) in biological systems *(1)*. The instrumental techniques used currently for NO measurements are spectroscopic and electrochemical methods *(2)*. Electrochemical methods offer several features that are not available from analytical spectroscopic methods. Most important is the capability afforded by the use of ultramicroelectrodes for direct in situ measurements of NO in single cells near the source of NO synthesis. NO released from the cell can be detected within a few miliseconds after injection of nitric synthase agonist, and the NO concentration on the membrane surface may vary from submicromolar to micromolar levels *(3)*.

NO is about seven times more soluble in the mambrane than in cytosol. Thus the membrane will be a storage reservoir for NO, and the small membrane volume can develop a relatively high concentration within a short period when NO is released by the NO-producing enzyme. From an analytical viewpoint, the detection of NO at the site of highest concentration is a convenient and accurate method of measurement of endogenous NO. Measurements of NO will be difficult after dilution in the aqueous phase in which concentrations are reduced, usually by a few orders of magnitude. Therefore, the placement of the electrochemical sensor on the membrane surface offers pratical advantages for the in situ monitoring of NO release. The measurement of NO from an isolated single cell requires a sensor to have a diameter smaller than the diameter of the cell.

The electrochemical determination of NO is based on its oxidation on conductive polymeric porphyrin electrode *(4)*. Generally, the oxidation of NO on

From: *Methods in Molecular Medicine, Vol. 36: Septic Shock*
Edited by: T. J. Evans © Humana Press Inc., Totowa, NJ

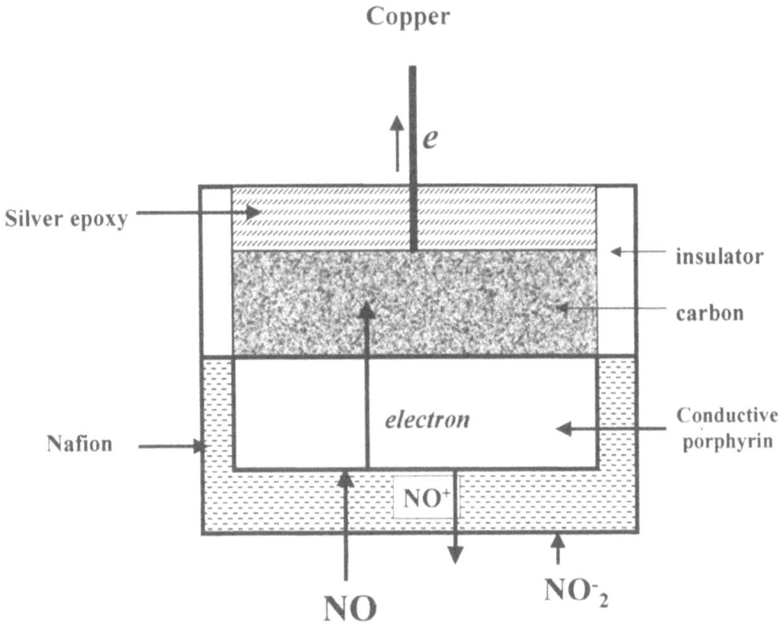

Fig. 1. Schematic diagram of porphyrinic sensor.

solid electrodes (gold, carbon, platinum) proceeds via two steps: electrochemical reaction 1 followed by a chemical reaction 2. The first electrochemical step is a one electron transfer from an NO molecule to the electrode resulting in the formation of a cation.

$$NO - e \rightarrow NO^+ \qquad (1)$$

NO^+ is a relatively strong Lewis acid and in the presence of OH^- is converted to nitrite (NO_2^-):

$$NO^+ + OH^- \rightarrow HNO_2 \qquad (2)$$

As the oxidation potential of nitrite in aqueous solution is only 60–80 mV more positive than that of NO, oxidation of NO on solid electrodes with scanned potential ends with two additional electrons transfered in with nitrite:

$$NO_2^- - 2e \rightarrow NO_3^- \qquad (3)$$

Figure 1 shows a porphyrinic sensor that has been developed for the electrochemical measurement of nitric oxide. The porphyrinic sensor is based on the electrochemical catalytic oxidation of NO on a conductive polymeric porphyrin (n-type semiconductor). The oxidation of NO on polymeric metalloporphyrin occurs at 600 mV (vs silver/silver chloride, SSCE). The current efficiency generated in this reaction is high, even at the physiologic pH 7.4 and current (analytical signal) is linearly proportional to NO concentration. To prevent reactions 2 and 3, a barrier made with cation exchanger polymeric Nafion is placed on the surface of polymeric porphyrin. The negatively charged layer of Nafion prevents the diffusion of NO_2^- from the bulk solution to a surface of the porphyrinic film. This eliminates potential interferences caused by nitrite ions that may be present in the analyzed system.

2. Materials and Instrumentation

2.1. Sensor Preparation

1. Carbon fibers (diameter 6–8 μm) specific resistivity 12 ohm cm (Amoco Performance Products, Grenville, SC).
2. Beeswax/rosin mixture (65:35 w/w).
3. Bare copper wire diameter 0.3–0.5 mm.
4. Glass electrolytic cell (19 × 48-mm^2 vial).
5. Teflon cap (19-mm diameter) with three holes (3-mm diameter each) and one hole (1-mm diameter).
6. Platinum auxiliary electrode (platinum wire, diameter 0.3–1 mm, 3-cm length).
7. Silver/silver chloride reference electrode (silver wire diameter 1–2 mm covered with thin layer of silver chloride)
8. Fast response (≤1 μs) computer controlled voltammetric analyzer (Princeton Applied Research, PAR 273, Princeton, NJ).
9. Nitrogen tank.
10. Propane microburner.
11. Capillary puller.
12. Glass capillaries (0.5 to 1-mm internal diameter).
13. Conductive silver epoxy glue (A.I.T., Lawrenceville, NJ).
14. Syringe needles (22-gage), 1-in length.
15. Chromatographic columns; florisil and silica gel.

2.2. Sensor Testing and Calibration

1. Porphyrinic sensor (working electrode).
2. Platinum auxiliary electrode.
3. Silver/silver chloride electrode.
4. Electrolytic cell.
5. Voltammetric analyzer.
6. Micromanipulator.
7. Nanopipet.
8. Conical reaction vial (5-mL) with 14/10 threaded screw cap with septum.
9. Nitric oxide generator or small (0.5–1 L) cylinder with gaseous NO.

2.3. Nitric Oxide Measurement

1. Porphyrinic sensor (working electrode).
2. Platinum auxiliary electrode.
3. Silver/silver chloride electrode (reference electrode).
4. Cell-culture dish.
5. Voltammetric analyzer.
6. Controlled micromanipulator (*x*,*y*,*z* resolution 0.2 μm).
7. Femtopipet.
8. Temperature-controlled incubator.
9. Faraday cage.

2.4. Reagents and Solutions

1. Monomeric tetrakis (3-methoxy-4-hydroxyphenyl) porphyrin Ni(II), (TMHPPNi) dissolved in 0.1 mol/L NaOH (final concentration 5×10^{-4} mol/L).
2. Sodium hydroxide, 0.1 mol/L.
3. Hanks balanced salt solution adjusted to pH 7.4 at 37°C (without phenol red) (Sigma, St Louis, MO).
4. Saturated solution of nitric oxide accumulated at 0°C in freshly boiled deionized, distilled water.
5. 1% Nafion obtained by dilution of 5% Nafion (Sigma) with 96% ethyl alcohol. Allow several hours for mixing after dilution.
6. Vanillin.
7. Pyrrole.
8. Propionic acid.
9. Dimethylformide.
10. Nickel (II) acetate tetraydrate.
11. Methylene chloride.
12. Methyl alcohol.

3. Methods

3.1. Preparation of Porphyrinic Sensor

3.1.1. Synthesis of Monomeric Porphyrin

1. TMHPPNi is obtained from vanillin and pyrrole (1:1 molar ratio) in boiling propionic acid (2–3 h).
2. The crude material, dissolved in minimal volume of methylene chloride, is first chromatographed on a florisil column using a mixture (1:1) of methylene chloride and methyl alcohol as an eluent; then chromatographed twice on a silica gel column using a gradient elution of methylene chloride/methyl alcohol gradually changed from 200:1 to 40:1.
3. Metallation of TMHPP is performed in dimethylforamide at 100°C with an excess of nickel(II) acetate tetrahydrate for 5 h (*see* **Note 1**).

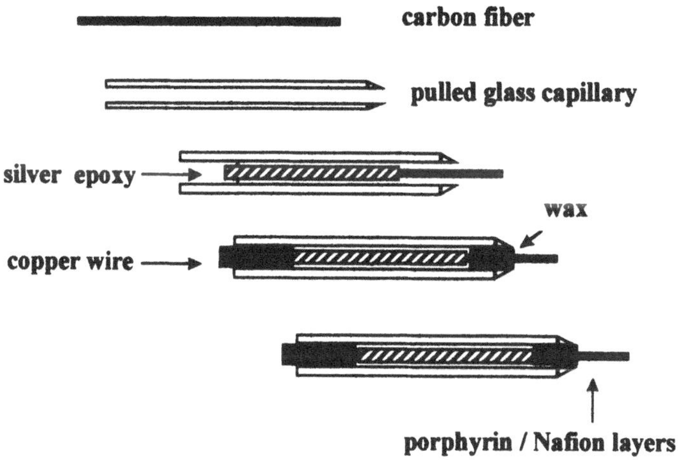

Fig. 2. Schematic diagram showing the preparation of a porphyrinic sensor.

3.1.2. Preparation of TMHPPNi Electrocoating Solution

1. Five milliliters of 0.1 mol/L NaOH is degassed for 15 min in a 2-dram (approx 10-mL) clear glass vial (electrolytic cell) by bubbling nitrogen delivered from a gas tank by 1- to 2-mm internal diameter glass pipet (1–2 bubbles/s for 15 min). After 15 min the pipet with continuously flowing nitrogen is raised above the meniscus and 1.5 mg of TMHPPNi is added.
2. The glass pipet with flowing gas is immersed again in solution (approx 5 mm below the meniscus) and the solution is stirred by the flowing nitrogen until all crystals of TMHPPNi are dissolved (approx 2 min).

3.1.3. Preparation of Carbon Fiber (see Note 2)

1. A 1.5 to 1.8-mm inner diameter open-ended capillary is pulled on one end to form a 50 to 100-μm diameter tip (capillary puller is used for this procedure).
2. The tapered tip is snipped off to produce a 10 to 15-μm opening (**Fig. 2**). A single carbon fiber (6–7 μm diameter) is threaded through the tapered end of the capillary with 1–2 cm of the fiber left protruding (*see* **Note 3**). The tapered glass tip–fiber electrode interface is sealed with a drop of beeswax/rosin mixture.
3. A copper wire contact is pulled to straighten out any bends. Then it is inserted through the untapered end of the capillary together with a small amount of silver epoxy. It is advanced until a conductive junction with the carbon fiber is made, as close as possible to the active tip. The capillary is then sealed with silver epoxy and the assembled electrode is transferred to a vacuum dryer (40°C) and dried for 12 h.
4. With the aid of a micromanipulator, the dry single fiber elctrode is thermally sharpened by placing it in close proximity to a propane microburner, where the

temperature is approx 1000°C. The position of the fiber should be perpendicular to the burner. A slow gradual burning will decrease the diameter of the elctrode's tip (sharpening process). The fiber should burn from its original length of 1–2 cm to 5–7 mm. Sharpening is required only if the sensor is going to be implanted in the cytoplasm of a cell. After the sharpening process, the fiber is immersed for 2–3 s in a molten mixture of beeswax and rosin (65:35 w/w) and sharpened again until the desired length and diameter are achieved. The active surface of electrode (that portion which is not covered with the wax) should have a length of 2–6 μm with a fiber diameter 0.4–2 μm, depending on sensor application. The smallest diameter tip which can be obtained in this process is 0.2 μm.

5. For measurement of NO release from several cells in cell culture or in tissue, a multifiber sensor can be used more effectively. The multifiber sensor is prepared according to the procedures described above with the omission of the burning and sharpening process. Multifiber sensors with diameters 15–30 μm (5–9 fibers) are more durable and easier to prepare. They also produce higher current, which makes measurment easier. For most applications, multifiber sensors will be sufficient.

3.1.4. Electrochemical Set-Up for Measurement of NO

1. The porphyrinic sensor is connected to any potentiostat for amperometric measurements, or to a voltammetric analyzer (potentiostat + waveform generator) for voltammetric measurements (*see* **Note 4**). A current sensitivity (for the instrument) of 100 pA/in will be sufficient for most measurements with multifiber (5–9 fiber) sensors. For a single-fiber small-diameter sensor, at least 10 times greater current sensitivity is required. This can be achieved by an equipping the potentiostat with a low noise current-sensitive preamplifier.
2. A three electrode system is used: a porphyrinic sensor (working electrode), a platinum wire (counter electrode), and a silver/silver chloride electrode. All three electrodes are connected to the potentiostat or voltammetric analyzer with low diameter (0.3–0.2 mm) copper wires shielded with copper mesh. For single-fiber measurements, all electrodes and the electrolytic cell and the biological matrix to be analyzed must be placed in a Faraday cage.
3. For amperometric measurements a constant potential of 0.68–0.70 V is applied. Current is monitored as a function of time and recorded with a chart recorder, x-y-t recorder, or computer with appropriate software. After recording the NO voltammogram or amperogram from a biological matrix, the calibration and a response of the sensor is verified by adding aliquots of saturated NO solution directly to the medium with biological sample (*see* **Note 5**).

3.1.5. Electrochemical Deposition of Polymeric TMHPPNi

1. Polymeric film is deposited electrochemically from an approx 5×10^{-4} mol/L solution of monomeric TMHPPNi in 5.0 mL of 0.1 mol/L NaOH contained in a 2-dram vial (electrolytic cell). Three electrodes: a sharpened carbon fiber elec-

trode (to be coated), a platinum wire counter electrode, and a reference SSCE are placed in the electrolytic cell through its Teflon cap with four holes (three for electrodes and one for a pipet to supply inert gas). After immersing the electrodes in the TMHPPNi solution, it is degassed again for 5 min. After the degassing process, the pipet (with inert gas flowing continuously) is raised 2–3 mm above the meniscus of the TMHPPNi solution. The electrodes are connected to a voltammetric analyzer and electrochemical deposition of polymeric TMHPPNi is initiated. The polymeric film is deposited by repeated cyclic scanning of the potential from –0.20–1.0 V and back with a potential scan rate of 100 mV/s (*see* **Note 6**). Usually 10–14 cycles are needed to provide sufficient coverage of the carbon fiber surface with polymeric TMHPPNi (*see* **Note 7**).

2. After deposition of polymeric film the electrode is rinsed with distilled water and transferred to another electrochemical cell with 5.0 mL of 0.1-mol/L NaOH and a single cyclic voltammogram is recorded in the range of potential –0.2–1.0V (vs SSCE) and scan rate 100 mV/s. The voltammogram should show two major peaks: an anodic peak at 0.54 V and cathodic peak at 0.40 V. When the peaks are not observed or very small, it is an indicator that polymeric TMHPPNi film was not deposited or that electrode coverage is insufficient.
3. The electrode is taken from NaOH solution and rinsed again with distilled water and allowed to dry in a vacuum oven at 40°C for 4 h.
4. The dry electrode is covered with Nafion by immersing it in a 1% solution in alcohol for 5–7 s, and then allowed it to dry. Usually 2–3 coats of Nafion are needed for sufficient Nafion coverage (*see* **Note 8**).

3.2. Calibration of the Sensor

1. A porphyrinic sensor, immersed in an electrolytic cell, can be calibrated by adding known volumes of aqueous solution saturated with NO (1.76 mmol/L at 0°C). A small tank (1–2 L) of pure nitric oxide (99.9%) is used to prepare standard solutions (*see* **Note 9**). A 5-ml vial is filled with phosphate buffer, then fitted with a septum, which is pierced with a 22-gage syringe needle to allow vapors to escape. After degassing by a brief boil (1 min), the phosphate buffer is chilled to 0°C and saturated with NO gas by bubbling it through the buffer (maximum time 10–15 min). NO is purified by passing two purge cylinders: the first filled with 2–3 mol/L NaOH and the second filled with distilled water. NO is delivered to the vial through the septum punched with an input 22-gage syringe needle. A hose attached to a second output needle of the same gage is used to dispose NO into the discharge air stream of a fume hood, or an aqueous solution of $KMnO_4$ (1 mol/L).
2. For sensor calibration, aliquots of saturated NO solution are injected (via gastight syringe) into the solution which will be used as a medium for biological experiments (*see* **Note 10**). An amperogram or voltammogram is measured after each injection of saturated NO standard. The current generated after each response is measured and plotted vs concentration. A linear relationship for cur-

rent vs concentration should be observed in the range of concentration 10^{-8}–10^{-5} mol/L (*see* **Note 11**).

3.3. Measurement of Nitric Oxide Release from Cells

1. Using a manual or motorized computer-controlled micromanipulator (0.2 µm *x,y,z* resolution), the porphyrinic sensor is implanted into a single cell, or placed on the surface of the cell membrane, or kept at a controlled distance (10 ± 2 µm) from an NO-generating cell. When the tip of the sensor touches the cell membrane, a transient small electrical noise is observed. This is a good indicator of zero distance from the cell and from this point the sensor can be moved out from the surface with 0.2-µm increments controlled by computer. A platinum counter electrode and reference silver/silver chloride electrode are placed in a solution at 0.5 to 1-cm distance from the sensor. All three electrodes are connected to voltammetric analyzer using copper wires (0.3-mm diameter). A constant potential is set at 0.65–0.70 V.
2. Injection of NO agonists is done with micro, nano, or femto injectors (*see* **Notes 12** and **13**).

4. Notes

1. Solid TMHPPNi is stable indefinitely, as long its kept away from excessive light, heat, or moisture. Degassed aqueous solutions of TMHPPNi are stable for at least 1 wk if kept refrigerated (4°C) and wrapped in aluminum foil, away from light, when not in use. Good solubility of TMHPPNi in 0.1-mol/L NaOH is an important indicator that the TMHPPNi is monomeric. Light and oxygen can initialize the formation of undesirable chemically polymerized TMHPPNi, which is poorly soluble in aqueous NaOH. This chemically polymerized TMHPPNi, if deposited on carbon, will show low electrical conductivity and catalytic properties for the oxidation of NO resulting in seriously decreased sensor sensitivity. Demetallation, or insufficient purification of TMHPPNi, may be additional reasons for poor performance of a sensor. Nickel (II) which is not coordinated to porphyrin (nickel hydroxide) can be deposited on the carbon fiber, and produce the Ni(II)/Ni(III) (peak Ia, Ic **Fig. 1**) couple on cyclic voltammogram. This kind of voltammogram can be assigned erroneously as a deposition of polymeric porphyrin.
2. The surface modification of a sharpened carbon fiber involves four steps: electrodeposition of a poly-TMHPPNi film on the electrode surface, conformation of poly-TMHPPNi film deposition, coverage with Nafion film, and testing an integrity of Nafion film. Each of these steps can influence the quality of the porphyrinic sensor.
3. The polymeric porphyrin has to be deposited on carbon with certain surface properties and electrical properties. The porphyrinic sensor cannot be prepared by using metal electronic conductors. Therefore, platinum/gold, iridium/platinum alloy wires cannot be used in sensor preparation. Carbon fibers with specific

resistivity 3–12 ohms/cm are used. The sensitivity of the sensor is slightly better when TMHPPNi is deposited on a low-resistivity fiber (3 ohms/cm). However, low-resistivity fibers with graphite-like structures are less flexible and more breakable than higher-resistivity fibers.

4. It is important to usc a high-sensitivy (picoamper level) and fast-response (microsecond) votammetric analyzer.
5. The quality of the sensor can assured by using differential pulse voltammetry. For differential pulse voltammetric measurements, the potential is scanned from 0.45–0.75 V with a scan rate of 2–5 mV/s. The pulse height can be set in the range of 20–40 mV and time interval can be set between 0.5–1 s. A differential pulse voltammogram (DPV) is a plot of current vs potential and can be recorded with an x-y recorder or computer. Depending on the pulse height and scan rate used in these measurements, as well as quality of the sensor, the peak current caused by the oxidation of NO should be observed in the range of potential of 0.63–0.65 V (vs SSCE). A peak potential higher than 0.70 V is indicative of a poor-quality sensor. Poor-quality sensors have both poor sensitivity, and poor selectivity.
6. Several features of continuous-scan voltammogram should be monitored during this process. Two voltammetric peaks should emerge after the first scan: one at 0.54 V caused by the oxidation of Ni(II) to Ni(III), the other at 0.40 V caused by the reduction of Ni(III) to Ni(II) (**Fig. 3A**). The peak current of both of these peaks should increase during each successive scan. Also, a high current should be observed at a potential 0.95–0.80 V. This current should increase and shift to a lower potential with each successive scan. The growth of the Ni(II)/Ni(III) peaks during each scan is an indicator of the amount of film being deposited. The high current observed at 0.95–0.80 V is caused by the oxidation of water to oxygen.
7. Overcoating (more than 20 scans under the condition described in this protocol) might decrease the sensitivity of the electrode significantly.
8. Nafion coverage and integrity of the porphyrinic film can be checked by cyclic voltammetry, in 0.1 *M* NaOH from –0.2–0.8 V. The absence of peaks at 0.54 and 0.40 V confirms adequate coverage of Nafion film (**Fig. 3B**). It is preferable to store the sensor in buffer solution (pH 7.4). The sensor can be stored dry for at least 6 mo without any loss of speed, sensitivity, or selectivity. The dry sensor can be sterilized with ethylene oxide. However, the sterilized dry sensor must be immersed in sterilized saline solution for at least 15 min before use.
9. NO is a toxic gas that reacts quickly with the oxygen in air. Therefore all operations with gaseous NO have to be performed in a well-ventilated fumehood using oxygen-free, inert-atmosphere, techniques. Aqueous solutions of NO can be stored inverted (septum down) for at least 48 h in a refrigerator (4°C).
10. The sensor is calibrated in static solution and all measurements are done under static conditions. Measurement of NO under laminar flow condition is possible, but technically difficult, requiring a special geometrical arrangement of the sensor as well as a special flow system for calibration and measurements *(5)*. The

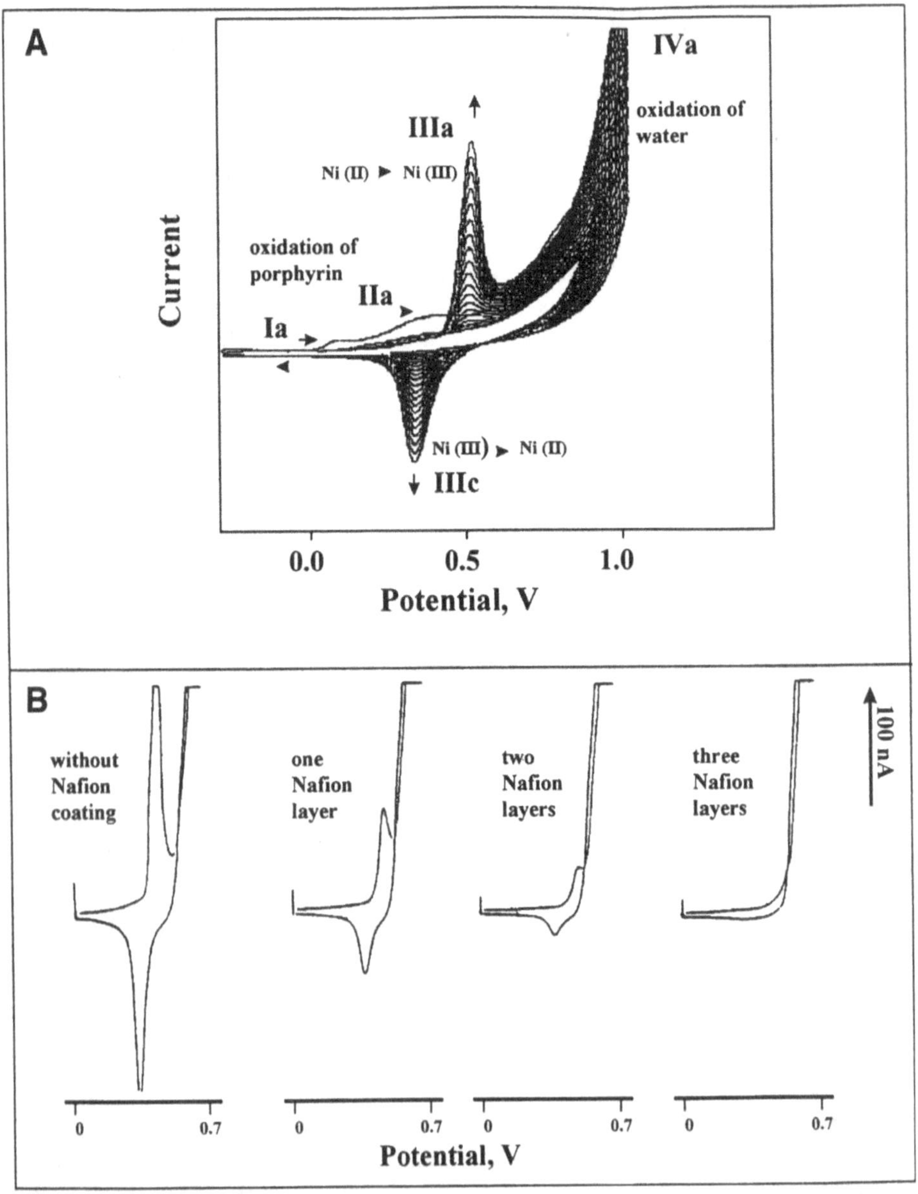

Fig. 3. Continuous scan cyclic voltammogram showing electrochemical deposition of polymeric TMHPPNi on carbon fiber (**A**); Single scan cyclic voltammograms obtained in 0.1-mol/L NaOH after sequential coating of polymeric-TMHPPNi with Nafion (**B**); Polymeric-TMHPPNi was coated by dipping electrode for 5 s in 1% Nafion and dried for 5 min.

description of this methodology is beyond the scope of this protocol. Under flow condition (blood perfusion) a rapid decrease of NO concentration is observed, therefore measured concentrations will be at least an order magnitude lower than that measured under static, diffusion-controlled conditions. The sensor cannot be used for quantitative measurement of NO in under turbulent flow conditions.

11. Calibration is done by preparing a calibration curve (usually in the range of 5×10^{-8} to 5×10^{-6} mol/L) or/and by the standard addition method. Crosschecking by using both methods is recommended. A standard calibration curve is prepared by subsequent addition of small volumes of standard aqueous saturated solution of NO into a constant volume of 0.1-mol/L phosphate buffer solution (pH 7.40 at 37°C). The current generated after each addition of NO is measured. The plot of current vs concentration should be linear. In the standard addition method, the addition of standard NO follows measurement of current caused by NO release from biological materials. It is extremely important that the sensor generates a high current per unit NO concentration. A typical current density for NO oxidation on porphyrinic sensor is between 0.3–1.8 mA cm^{-2} μmol^{-1} L and at least four to seven times higher than that which can be obtained on activated carbon fiber covered with Nafion. A current density for NO oxidation on porphyrinic sensor depends on film quality and should be at least 0.4 mA cm^{-2} μmol^{-1} L to have a sensor applicable for measurements in biological environments. The sensor can show a slightly different response depending on the composition of the medium (matrix effect). The pH dye (phenol red) used in some culture media can be strongly absorbed on a surface of the sensor, and can affect its performance. Therefore use of the cell- or tissue-culture media with pH dyes should be avoided.
12. Injection of larger volumes of agonist with a micro injector to a vial with biological material will cause a "jet effect" and the first release of NO will be caused by shear stress *(5,6)*. The shear stress peak of NO will be followed by a NO peak due to nitric oxide agonist. The time resolution of these two peaks can be small therefore, they may be observed only with a high-speed recording device (computer-controlled acquisition system), with chart recorders these two peaks will be indistinguishable. Injection of the agonist with a nano or femto injector will reduce the jet effect and subsequent NO release caused by shear stress. However, the concentration of the agonist will be lower than that injected, and the response time to the agonist will be longer. For most studies that do not involve studies of the kinetics (pattern) of NO release, the injection of agonist with micro or nano injector will be preferable.
13. The porphyrinic sensor described here is to fragile to be implanted in the deep tissue. For this purpose a catheter protected sensor of different design has to be used *(7,8)*.

Acknowledgments

This work was supported by a grant from the U.S. Public Health Service (HL-55397) and Research Excellence Fund, Center for Biomedical Research, Oakland University.

References

1. Kiechle, F. and Malinski, T. (1993) Nitric oxide: biochemistry, pathophysiology and detection *Am. J. Clinical Pathology* **100,** 567–575.
2. Malinski, T., Kubaszewski, E., and Kiechle, F. (1996) Electrochemical and spectroscopic methods of nitric oxide detection. *Methods Neurosci.* **31,** 14–33.
3. Malinski, T., Taha, Z., Grunfeld, S., Patton, S., Kapturczak, M., and Tomboulian, P. (1993) Diffusion of nitric-oxide in the aorta wall monitored in-situ by porphyrinic microsensors. *Biochem. Biophys. Res. Comm.* **193,** 1076–1082.
4. Malinski, T. and Taha, Z. (1992) Nitric oxide release from a single cell measured in situ by a porphyrinic microsensor. *Nature* **358,** 676–678.
5. Kanai, A. J., Strauss, H. C., Truskey, G. A., Crews, A. L., Grunfeld, S., and Malinski, T. (1995) Shear stress induces ATP-independent transient nitric oxide release from vascular endothelial cells, measured directly with a porphyrinic microsensor. *Circ. Res.* **77,** 284–293.
6. Malinski, T. and Czuchajowski, L. (1996) Nitric oxide measurements by electrochemical methods, in *Methods in Nitric Oxide Research* (Feelisch, M. and Stamler, J., eds.) John Wiley & Sons, Chichester, UK, pp. 319–339.
7. Vallance, P., Patton, S., Bhagat, K., MacAllister, R., Radomski, M., Moncada, S., and Malinski, T. (1995) Direct measurement of nitric oxide in human beings. *Lancet* **345,** 153–154.
8. Pinsky, D., Patton, S., Mesaros, S., Brovkovych, V., Kubaszewski, E., Grunfeld, S., and Malinski, T. (1997) Mechanical transduction of nitric oxide synthesis in the beating heart. *Circ. Res.* **81,** 372–379.

13

Immunochemical Detection of Nitric Oxide Synthase in Human Tissue

Lee D. K. Buttery and Julia M. Polak

1. Introduction

To date three distinct isoforms of nitric oxide synthase (NOS) have been identified. Two isoforms are considered to be expressed constitutively—neuronal NOS (nNOS; type I NOS) and endothelial NOS (eNOS; type III NOS). The third isoform is not generally present in normal cells and tissues but is induced in response to infection, inflammation or trauma—inducible NOS (iNOS; type II NOS). In 1990 Bredt and Synder *(1)* succeeded in developing antibodies to rat brain NOS (nNOS) and used immuncytochemistry subsequently to furnish one of the first anatomical descriptions of the distribution and localization of nNOS. Today numerous antibodies to all three NOS isoforms isolated from various tissues and different animal species are available and the application of immunocytochemistry is commonplace in the investigation of NOS in healthy and diseased tissues including human *(2–6)* (**Figs. 1–3**).

Immunocytochemistry originally developed in the 1940s is now a proven technique for identifying and ascertaining the detailed localization and distribution of specific proteins within cells and tissues *(7)*. There are numerous modifications of the basic method and here we will discuss primarily the use of the avidin-biotin-peroxidase complex (ABC) method and the indirect immunofluorescence method to identify either single or multiple antigens in both cultured cells and tissue sections. Although the concept of immunocytochemistry and the methodology are relatively simple there are several important criteria that must be considered for successful identification and visualization of antigens within cells and tissues. Perhaps of paramount importance is efficient preservation of the cells and tissues being investigated. Tissue fixation serves

From: *Methods in Molecular Medicine, Vol. 36: Septic Shock*
Edited by: T. J. Evans © Humana Press Inc., Totowa, NJ

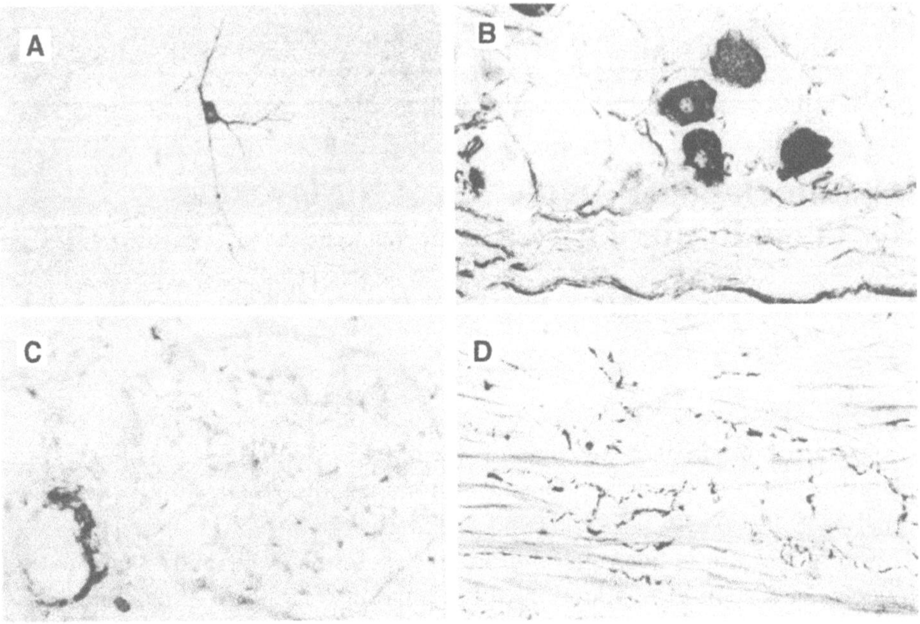

Fig. 1. Immunostaining for nNOS in (**A**) multipolar neurons in human cerebral cortex, (**B**) human dorsal root ganglia, (**C**) macula densa in human kidney, (**D**) nerve fibres supplying human tracheal smooth muscle.

two purposes: preservation of tissue structure and preservation of the antigen of interest, in situ, without significant loss of antigenicity. Again there are numerous approaches to this subject but here we focus on the use of paraformaldehye fixatives. Once the cells or tissue have been fixed they need to be processed appropriately involving washing, and if blocks of tissue are being used, cutting of thin tissue sections. Here we focus primarily on preparation of cryostat blocks and cutting of frozen sections.

Installation of rigorous specificity controls into each and every immunocytochemical method is also essential to achieve consistent and clean identification of specific antigens and to instil confidence that the protein being investigated is indeed the one of interest. Whereas we discuss the inclusion of such specificity controls, it is often desirable to also perform Western blotting (not discussed in this chapter) on the tissue type and species being investigated to ensure/confirm the antibody characteristics. However, when carefully controlled, immunocytochemistry is a powerful technique producing meaningful data on the presence and distribution of specific proteins within tissues and is also an efficient means of quantifying potential disease-related changes in their expression and distribution.

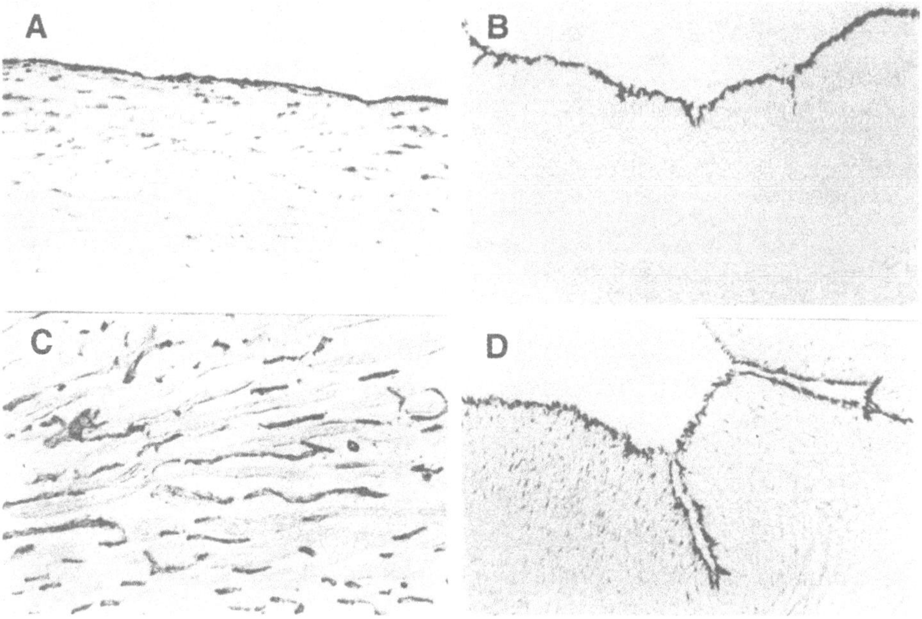

Fig. 2. Immunostaining for eNOS in **(A)** human aorta, **(B)** human coronary artery, **(C)** intra-myocardial capillaries in human heart, **(D)** human umbilical vein.

2. Materials

2.1. General

2.1.1. Basic Equipment

The basic equipment required for immunocytochemistry includes metal staining racks (Histological Equipment, Nottingham, UK) that are useful for washing and staining procedures; plastic storage containers (approx 500-mL capacity) that are useful for washing and staining procedures; and plastic Petri dishes (approx 20-cm diameter) that are useful for antibody incubation procedures. If sectioning tissue, access to cryostat (or wax microtome) is required. However, many institutions offer a tissue-section cutting service. Access to a photo microscope (bright field and fluorescence) is also necessary.

2.1.2. Antibodies to NOS

For the most part, we have used NOS antibodies raised "in house" *(8–13)*. This is not because these antibodies are better, but because until recently, commercial antibodies were not readily available. Today numerous polyclonal and monoclonal antibodies to all NOS isoforms (extracts/peptides, N-terminal/

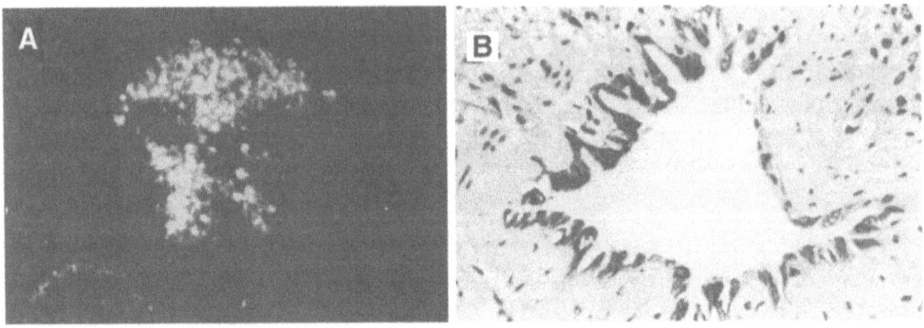

Fig. 3. Immunostaining for iNOS in **(A)** primary granules in cytokine-stimulated human neutrophils, **(B)** bronchiolar epithelium of diseased human lung.

C-terminal) are available commercially from various reputable suppliers, including Alexis Biochemicals (Alexis Corporation, Nottingham, UK; www.alexis-corp.com), Santa Cruz Biotechnology (Santa Cruz, USA; www.scbt.com), Transduction Laboratories (Affiniti Research Products, Exeter, UK; www.affiniti-res.com), and Sigma (Poole, UK).

It not our intention to recommend any one particular antibody. In our experience and also that of other workers ***(14)***, NOS antibodies should be evaluated by the individual user on the particular tissue of interest and to also take into account the effects of any experimental conditions/agents to which the tissues might have been exposed. In addition to the specificity controls detailed in this chapter (**Subheading 3.6.**), it is advisable to also use Western blotting to further evaluate the characteristics of the antibody. Potential specificity problems might include, for example, crossreaction with other NOS isoforms. Because of the high interspecies conservation between specific NOS isoforms, NOS antibodies do demonstrate cross-species reactivity. However, it is possible that unexpected localizations might arise using such antibodies and in some instances it may be appropriate to use antibodies common to the species and even tissue under investigation ***(14)***.

2.2. Buffers/Antibody Diluents

1. Phosphate-buffered saline (PBS; 10 m*M*): To 1 L of deionized water add 8.79 g NaCl, 0.272 g KH_2PO_4, 1.135 g Na_2HPO_4, mix, and dissolve thoroughly. Adjust to pH 7.2–7.4 with concentrated HCl. Store at room temperature. Prepare fresh each week (*see* **Note 1**).
2. Diluent for primary antibodies and normal sera: PBS, 0.05% bovine serum albumin (BSA), 0.01% NaN_3 (PBS:BSA:NaN_3). In 100 mL of 10 m*M* PBS dissolve 0.05 g BSA and 0.01g NaN_3. Store at 4°C. Stable for several weeks (*see* **Note 2**).

3. Diluent for secondary antibodies: PBS:BSA; In 100 mL of 10 m*M* PBS dissolve 0.05 g BSA. Sodium azide is not added as during the course of detection of bound antibody it can interfere with the precipitation of chromogens like diaminobenzidine. Store at 4°C. Prepare fresh each week.

2.3. Tissue Preparation

1. Paraformaldehyde –1% solution (w/v): Heat 1 L of 10 m*M* PBS to 60°C and add 10 g of paraformaldehyde and stir until dissolved thoroughly and allow to cool. Prepare fresh or store in aliquots at –20°C. Toxic by inhalation and contact with skin (*see* **Note 3**).
2. PBS:sucrose washing buffer: Dissolve 150 g of sucrose in 1 L of 10 m*M* PBS and add 10 mg of sodium azide (NaN_3) to inhibit microbial/fungal contamination. Store at room temperature. Stable for many weeks.

2.4. Tissue Pretreatment

1. Methanol:hydrogen peroxide blocking buffer: 0.03% (v/v) hydrogen peroxide (30% solution) in methanol. Toxic by inhalation and skin contact. Prepare fresh.
2. PBS:Triton: 0.2% (v/v) Triton X-100 in 10 m*M* PBS. Prepare fresh each week. Store at room temperature.
3. Pontamine sky blue: In 300 mL of 10 m*M* PBS dissolve 0.15 g pontamine sky blue and add 1.5 mL dimethyl sulphoxide. Filter (Whatman 113 V) and store a room temperature. Prepare fresh each week.

2.5. Detection of Bound Antibodies (*see* Note 4)

1. 3',3'-Diaminobenzidine tetra-hydrochloride (DAB): In 200 mL of PBS dissolve 5 g of DAB (Sigma cat. no. D5637). Once dissolved, DAB forms a brown or purple solution. Filter (Whatman 113V) and store in 4-mL aliquots at –20°C. DAB is a suspected carcinogen.
2. Glucose oxidase–nickel enhancement of DAB:
   a. Acetate buffer (0.1 *M*): To 1 L of deionized water add 13.61 g sodium acetate, dissolve and adjust to pH 6.0 with glacial acetic acid. Store at room temperature. Stable for several weeks.
   b. DAB enhancement buffer: To 400 mL of 0.1 *M* acetate buffer add 10 g nickel ammonium sulphate, 160 mg ammonium chloride, and 800 mg D-glucose and allow to dissolve thoroughly. To this add one 4-mL aliquot of DAB and 12 mg glucose oxidase (Sigma, Type VII, cat. no. G1233); stir to dissolve completely. Prepare fresh. Use immediately.

3. Methods

3.1. Preparation of Antibodies

1. Antibodies should be stored in aliquots (–40°C) or according to manufacturer's instructions. The toxicity of many antibodies is unknown, avoid skin contact (*see* **Note 5**).

2. Antibody aliquots are defrosted and optimally diluted in PBS:BSA:NaN_3 and where possible prepared fresh.

3.2. Tissue Preparation

1. Fix tissue of interest in a 1% solution of paraformaldehyde for 6–8 h at room temperature or overnight at 4°C (*see* **Note 6**).
2. After fixing, tissues are washed thoroughly in several changes of PBS:sucrose over 2–3 d.
3. Cryostat blocks are prepared by orientating tissues on a cork tile (approx 1-cm diameter), surrounding it in mounting medium (Tissue Tek, Miles, Elkhart, IN; Raymond A. Lamb, Eastbourne, UK), and freezing rapidly by submersion into a beaker containing melting isopentane, previously frozen by partially submersing the beaker in liquid nitrogen (*see* **Note 7**).
4. Section tissues (5–10 μ*M*) in cryostat at –25°C and thaw-mount tissue sections on to treated slides and leave to dry for 1 h at room temperature (*see* **Note 8**).

3.3. ABC Method

1. Rehydrate air-dried tissue sections by immersion in 10 m*M* PBS for 5 min.
2. Block endogenous peroxidase by immersion in methanol:hydrogen peroxide blocking buffer for 20 min at room temperature.
3. Wash slides in fresh changes of 10 m*M* PBS (3× 5-min each).
4. Block nonspecific binding sites by incubating sections in a 3% solution (v/v) of normal serum in PBS:BSA:NaN_3 for 20 min at room temperature (*see* **Note 9**).
5. Blot excess normal serum from the slide and replace with primary antibody optimally diluted in PBS:BSA:NaN_3 and incubate overnight at 4°C in a moistened chamber (*see* **Note 10**).
6. Wash slides in fresh changes of PBS (3× 5 min).
7. Incubate sections with biotinylated secondary antibodies, specific (species and immunoglobulin subtype) for the primary antibody for 30 min at room temperature (*see* **Note 11**).
8. Wash slides in fresh changes of PBS (3× 5 min).
9. Incubate sections in ABC reagent (Vectastain, Vector Laboratories, Peterborough, UK) for 1 h.
10. Wash slides in fresh changes of PBS (3× 5 min).
11. To develop antibody-labeled sections (*see* **Note 12**) dissolve DAB solution in 10 m*M* PBS (1× 4-mL aliquot per 400 mL PBS), add 16–20 drops of hyrogen peroxide (30% solution) using a 1-mL pipet and stir. Incubate sections for 30 s to 5 min, checking color development (brown deposit) microscopically.
12. Alternatively, sections can be developed according to the glucose oxidase–DAB–nickel enhancement method. Sections are allowed to equilibrate in fresh changes of acetate buffer (2x 5 min) and sections are then incubated in DAB enhancement buffer for 3–5 min, checking color development (blue-black deposit) microscopically.

13. Wash in fresh changes of PBS (1x 5 min) and then tap water 2× 5 min) (*see* **Note 13**).
14. Dehydrate through a graded ethanol series (70, 90, 99, 99%), clear (xylene or similar agent) and mount (Pertex or similar compound).

3.4. Immunofluorescence (*see* Note 14)

1. Rehydrate air-dried sections by immersion in 10 m*M* PBS for 5 min.
2. Permeabilize sections for 45 min in PBS:Triton.
3. Wash slides in fresh changes of 10 m*M* PBS (3× 5 min).
4. Counterstain in Pontamine sky blue for 15–20 min.
5. Wash in fresh changes of 10 m*M* PBS (3× 5 min).
6. Block nonspecific binding by incubating for 20 min in a 3% (v/v) solution of normal serum in PBS:BSA:NaN_3.
7. Blot the excess serum and apply primary antibody optimally diluted in PBS:BSA:NaN_3 and incubate overnight at 4°C in a moistened chamber.
8. Wash slides in fresh changes of 10 m*M* PBS (3x 5 min).
9. Incubate sections with fluorescein isothiocyanate (FITC)-conjugated secondary antibody for 1 h at room temperature. Alternative conjugates can be used including tetra-rhodamine isothiocyanate (TRITC) or Texas Red.
10. Wash slides in fresh changes of 10 m*M* PBS (3x 5 min).
11. Mount in 1:1(v/v) PBS:glycerol or Vectashield (Vector Laboratories), which contains an antifade agent, prolonging fluorescence lifetime.

3.5. Double Immunofluorescence Immunostaining (*see* Note 1 5)

1. Perform **steps 1–5** in **Subheading 3.4.**
2. Blot excess serum and apply the first primary antibody (i.e., monoclonal) optimally diluted in PBS:BSA:NaN_3 and incubate overnight at 4°C in moistened chamber.
3. Wash in fresh changes of 10 m*M* PBS (3× 5 min).
4. Incubate sections with FITC-conjugated secondary antibody (i.e., horse anti-mouse) for 1 h at room temperature.
5. Wash in fresh changes of 10 m*M* PBS.
6. Apply second primary antibody (i.e., rabbit polyclonal) optimally diluted in PBS:BSA:NaN_3 and incubate overnight at 4°C in moistened chamber. Prior to incubating in primary antibody it is advisable to block nonspecific binding by re-incubating in normal serum.
7. Wash in fresh changes of 10 m*M* PBS (3× 5 min).
8. Incubate sections with TRITC-conjugtaed secondary antibody (i.e., goat anti-rabbit) for 1 h at room temperature.
9. Wash in fresh changes of 10 m*M* PBS (3× 5 min).
10. Mount in 1:1 (v/v) 10 m*M* PBS:glycerol or Vectashield.

3.6. Specificity Controls

In addition to the various controls detailed in this chapter, for example, inclusion of normal, nonimmune sera and antibody titration, it is imperative to include positive, negative, and experimental controls into every immuncytochemical procedure.

1. Negative controls are performed to ensure that the tissue investigated is immunostained specifically. For polyclonals, nonimmune (or ideally preimmune serum) is substituted (appropriately titrated) for the primary antiserum. For monclonals an inappropriate antibody (i.e., one not expected to be present) is used. The secondary antibodies are evaluated by simply omitting the primary antibodies—no staining should be seen.
2. Positive controls (tissues known to contain the antigen of interest) should be included to evaluate a previously untested antibody and ideally should be included in every immunocytochemical procedure as a qualitative assessment of the antibody.
3. Experimental controls typically involve quenching of immunostaining by previous absorption of the antibody with antigen. Many antibodies to NOS are now raised to peptide fragments and these peptides (usually available from the supplier) can readily be used for quenching experiments. Antigen is diluted typically to a final concentration of 10, 1, 0.1, 0.01, 0.001 nmol/mL and mixed with primary antibody, optimally diluted for approx 60 min at room temperature. Thereafter, sections are incubated with antigen-antibody mixture according to the normal protocol. With increasing concentration of antigen staining should disappear or be "quenched." This is performed in parallel with a positive control. The procedure should be repeated using an unrelated antigen to quench the antibody of interest or using the specific antigen to quench an unrelated antibody. In both situations staining should be unaffected.

3.7. Microscopy

1. Sections stained by the ABC-peroxidase method are viewed using a conventional bright-field microscope fitted with a (blue) daylight filter. If the microscope is fitted with Normarski or differential interference optics they can be used to enhance image contrast producing a pseudo-three-dimensional effect with cell/tissue structures seen in relief. This is particularly useful if counterstaining has not been used. A good film for color exposures is Fujichrome Provia 100 daylight color-reversal film. For black and white photography we routinely use Kodak T-Max 100 (negative) film.
2. Sections stained by immunofluorescence methods should be photographed immediately as the signal will tend to fade, particularly if the stained preparation is viewed repeatedly or for extended periods of time. Most fluorescence microscopes are fitted with appropriate filters to excite conventional fluorophores (FITC/TRITC/Texas Red). For simultaneous visualization of double immuno-

fluorescence double-pass filter blocks are available. For color exposures we routinely use Fujichrome 400 daylight film. For black and white photography Kodak Tri-X 400 is generally suitable.

4. Notes

1. Alternative buffers are available, notably Tris-buffered saline (0.05 *M*) at pH 7.6, which is often used in alkaline phosphatase immunocytochemistry.
2. Triton X-100 (0.2%, v/v) can also be added to antibody diluent and helps to increase tissue permeabilization, improving antibody penetration.
3. We have found a 1% solution of paraformaldehyde to be a good general fixative for NOS immunocytochemistry in both dissected tissues and cultured cells. A higher concentration of paraformaldehyde (4% solution, w/v) can also be used and is particularly useful for investigation of central nervous system tissues. If frozen aliquots of fixative are used, a white precipitate may form on thawing which will redissolve by warming to approx 40°C. In our experience other fixatives such as Zamboni's and Bouin's, although perfectly adequate for tissue preservation, do not yield as good results compared to paraformaldehyde for NOS immunocytochemistry.
4. Many of the reagents used in revealing bound antibodies are toxic or suspected carcinogens, in particular DAB and nickel ammonium sulphate. Gloves should be worn when handling or preparing any of these solutions.
5. When using an antibody for the first time or testing on an unknown tissue, the antibody should be titrated to determine an optimal dilution. Ideally, a known positive control tissue should be used. For nNOS rat/human brain/gut is suitable. For eNOS normal, intact, artery/vein should be used (placental tissue is suitable and is often easy to obtain). For iNOS the situation is more difficult but cytokine-challenged murine RAW macrophages/human peripheral blood neutrophils or endotoxin-treated rat tissues are generally suitable.

   Optimal dilution of an antibody is defined as the highest at which specific immunoglobulin can saturate the available antigen. To some extent this is a subjective process but the aim is to determine a dilution at which, once the antibody binding sites are revealed, there is good contrast between staining of specific cells/tissue compartments and the rest of the tissue (background). In a typical titration experiment, a series of antibody dilutions is made—1/50, 1/100, 1/200, 1/500, 1/1000, 1/2000, 1/5000, 1/10,000, and so on. Staining with antibody is then graded using an arbitrary scale; 0 no staining, + weak staining, ++ moderate staining, and +++ strong staining. This scale is used to grade both specific and background staining.

   Of course temperature, time of incubation, and dilution of secondary antibody will affect antibody dilution (*see* **Note 12**). Moreover, sensitivity of the technique is important. As a general rule of thumb, with two-step indirect methods (i.e., **Subheadings 3.4.** and **3.5.**—immunofluorescence) the antibody dilution is 5–10 times *lower* as compared to a three step method (i.e., **Subheading 3.3.** ABC–peroxidase).

6. Good preparation of tissue is essential to clear immunocytochemical demonstration of proteins in situ. Where possible we collect tissue directly at the time of surgery (transporting in sterile 10 m*M* PBS on wet ice). To ensure adequate tissue fixation we generally dissect samples to sizes no greater than 1 × 1 × 0.5 cm and immerse the tissue in fixative (frozen aliquots of fixative should be allowed to warm to room temperature). The ratio of fixative to tissue should be approx 4:1. Larger tissue pieces can be used but fixation times will have to be increased. If whole organs or large pieces of tissue are available it is often advisable is locate a blood vessel and carefully perfuse or distend (in the case of airways in lung tissue) with fixative via a syringe fitted with blunt needle. Although this helps distribute fixative throughout the tissue, it is still advisable to also dissect the tissue and perform immersion fixation.

   In the case of cell cultures, media is decanted, cells washed with 10 m*M* PBS, and cultures incubated in fixative for 20–30 min.
7. Good preparation of blocks can alleviate potential problems with sectioning. For larger pieces of tissue (1 × 1 cm) we often prepare a mold made from autoclave tape (wrapped around the cork tile) and this produces an evenly shaped block and minimizes formation of air bubbles. For transverse orientation of blood vessels, a small pin can often be inserted carefully down the lumen to prevent its moving when mounting medium is poured. Also, when sectioning tissue, particularly human tissues, gloves should be worn to minimize potential risks of infection.

   Although we have described a method for cryostat sections it is possible to use paraffin-embedded tissues and sections. Clearly before use these sections have to be dewaxed by immersion in xylene (or similar solvent) and rehydrated through a graded ethanol series (90%, 90%, 70%, 50%) to water. Results of NOS immunocytochemistry on paraffin sections are generally poor unless an antigen-retrieval step is performed. We have found autoclaving sections in 0.01 *M* citrate buffer at pH 6.0 markedly improves NOS immunocytochemistry. If using a small bench-top autoclave two cycles is recommended ***(15)***.
8. Glass slides are treated to help improve section adhesion and reduce the risk of the section becoming detached and floating away during the staining procedure. We usually treat slides with Vectabond (Vector Laboratories) which chemically damages the surface of the glass slide. Poly-L-lysine (molecular mass 150,000–300,000) coated slides can also be used. However, we often find that nonspecific staining is increased. This is often particularly prominent at the margins of the tissue and in the lumen of blood vessels/airways and is therefore problematic for investigating staining in these regions of the tissue.
9. Inclusion of normal nonimmune serum helps to prevent nonspecific binding of the second (label-conjugated) antibody. The choice of serum is dependent on what species the second antibody is raised. Common species include goat, horse, and rabbit. Thus, if using a goat anti-rabbit secondary antibody normal goat serum would be the blocking serum.
10. We routinely incubate sections in primary antibody overnight at 4°C but it is also possible to perform incubations at room temperature or even at 37°C. Although

increasing the temperature reduces incubation times (to approx 1 h), we have found that there is often an increase in background-staining intensity. Remember also that if changing from one incubation regime to another, the antibody should be retitrated under the new conditions.

11. With regard to dilution of secondary antibodies, we aim to select an optimal dilution, which then generally remains static even if other conditions are altered. Often the supplier (Vector Laboratories; Sigma) will enclose information on the antibody characteristics including a suggested dilution. As a general rule secondary antibodies are diluted 1/100. Normal serum (2%, v/v) from the same species as the tissue being investigated (i.e., human) can be included in the secondary antibody and this can help to limit problems encountered with background staining and particularly those that are not remedied by altering antibody dilutions.
12. The DAB development solution can remain active for approx 1 h and may be used several times within this period. After use, DAB solution is discarded by first adding a few drops of bleach which inactivates/oxidizes DAB (solution turns black) and then flushing down the sink with plenty of water.
13. Before mounting, sections may be counterstained to provide contrast between the immunostaining and the background and to also give structure to the tissue as a whole. If DAB alone is used as chromogen, hematoxylin is used as counterstain. Slides are immersed in hematoxylin (i.e., Harris's Haematoxylin, BDH, Poole, UK) for 30–60 s and washed in running tap water for approx 10 min or until the blue color develops (this can be shortened if the water is hand warm). Counterstaining is differentiated by immersing (approx 5 s) in acid-alcohol (1% HCL in 70% alcohol) and then washing in tap water. If DAB staining is enhanced by nickel then the sections should be counterstained for 60 s in a 1% solution of neutral red or methyl green. Because these dyes are very water-soluble, sections are dipped only very briefly in water and then carefully blotted with filter paper, before washing in 99% ethanol, xylene and mounting.
14. We have given a general method for immunofluorescence. Although the ABC-peroxidase method is perfectly adequate—and in some instances is preferable to immunofluorescence as the stained preparations generally do not fade—the immunofluorescence method has the advantage that it can be more readily adapted for further applications. These include examination of thick tissue sections and demonstration of tortuous structures such as nerves by confocal microscopy or colocalization studies (*see* **Subheading 3.5.**). Remember also from **Note 5** that immunofluorescence is a two-stage method and thus the antibody is used at higher concentrations. However, some antibodies do not dilute well and therefore may not be suitable for immunofluorescence (i.e., it becomes very costly if antibodies are used at high concentrations).

    Immunofluorescence staining of thick tissue sections (50–100 μm) is performed with the section "free-floating" in a small Petridish and all solutions are added to and decanted from the section. Overall, however, the same routine (*see* **Subheading 3.4.**) is followed except incubation times (depending on section thickness) have to be increased; **Subheading 3.4., step 2:** incubate for up to 4 hr;

Subheading 3.4., step 6: incubate for up to 60 min; **Subheading 3.4., step 7:** incubate for up to 48 hr; **Subheading 3.4., step 9:** incubate for up to 3 hr. All washing stages should be increased to approx 30 min.

15. Fluorophores (i.e., FITC and TRITC) have distinct excitation and emission profiles and are therefore useful for colocalization studies, as there is little or no extinction/masking of signal from the separate bound antibodies. For colocalization using the method described here separate monoclonal and polyclonal antibodies (i.e., antibodies raised in different species) are used. Moreover, the species-specific separately labelled secondary antibodies must be noncrossreactive with each other (i.e., raised in or against separate species). We generally use a 3-d staining procedure incubating and detecting the antibodies individually but it is possible to apply the antibodies and perform incubations simultaneously. If using the longer method the antibody known to give the strongest signal should be applied first.

    It is possible to perform double immunostaining using the ABC method (or a combination of ABC and alkaline phosphatase). For example, with the ABC method the first antibody is developed using the DAB-nickel enhancement approach. Sections are then incubated with the second primary antibody and developed using DAB alone. This gives separate blue-black and brown signals (alternative chromogen/substrate kits are available from Vector Laboratories). However, we only recommend this method if the antigens of interest are in separate cells or tissue compartments to avoid masking of signal. Use of antibody elution and antibodies from the same species is also possible for multiple labeling but is beyond the scope of this chapter.

References

1. Bredt, D. S., Hwang, P. M., and Snyder, S. H. (1990) Localization of nitric oxide synthase indicating a neural role for nitric oxide. *Nature* **347,** 768–770.
2. Springall, D. R., Riveros-Moreno, V., Buttery, L. D. K., Suburo, A., Bishop, A. E., Moncada, S., and Polak, J. M. (1990) Immunological detection of nitric oxide synthase(s) in human tissues using heterologous antibodies suggesting different isoforms. *Histochemistry* **98,** 259–266.
3. Kobzick, L., Bredt, D. S., Lowenstein, C. J., Drazen, J., Gaston, B., Sugarbaker, D., and Stamler, J. S. (1993) Nitric oxide synthase in human and rat lung: immunocytochemical and histochemical localization. *J. Histochem. Cytochem.* **9,** 371–377.
4 Shaul, P. W., North, A. J., Wu, L. C., Wells, L. B., Brannon, T. S., Lau, K. S., Michel, T., Margraf, L. R., amd Star, R. A. (1994) Endothelial nitic oxide synthase is expressed in cultured human bronchiolar epithelium. *J. Clin. Invest.* **94,** 2231–2236.
5. Tracey, W. R., Xue, C., Klinhhofer, V., Barlow, J., Pollock, J. S., Forstermann, U., and Johns, R. A. (1994) Immunological detection of inducible NO synthase in human lung. *Am. J. Physiol.* **266,** L722–727.

6. Nicholson, S., Bonecini-Almeida, Lapa e Silva, J. R., Nathan, C., Kie, Q. W., Mumford, R., Weidner, R., Calagcay, J., Geng, J., Boechat, N., Linhares, C., Rom, W., and Ho, J. L. (1996) Inducible nitric oxide synthase in pulmonary alveolar macrophages from patients with tuberculosis. *J. Exp. Med.* **183,** 2293–2302.
7. Polak, J. M. and Van Noorden, S., eds. (1997) *Introduction to Immuncytochemistry*, BIOS Scientific Publishers, Oxford, UK.
8. Pollock, J. S., Nakane, M., Buttery, L. D. K., Martinez, A., Springall, D. R., Polak, J. M., Forstermann, U., and Murad, F. (1993) Characterization and localization of endothelial nitric oxide synthase using specific monoclonal antibodies. *Am. J. Physiol.* **265,** C1379–C1387.
9. Buttery, L. D. K., McCarthy, A., Springall, D. R., Sullivan, M. H. F., Elder, M. G., Michel, T., and Polak, J. M. (1994) Endothelial nitric oxide synthase in the human placenta: regional distribution and proposed regulatory role at the feto-maternal interface. *Placenta*, **15,** 257–265.
10. Buttery, L. D. K., Chester, A. H., Springall, D. R., Borland, J. A. A., Michel, T., Yacoub, M. H., and Polak, J. M. (1996) Explanted vein grafts with an intact endothelium demonstrate reduced focal expression of endothelial NO synthase specific to atherosclerotic sites. *J. Pathol.* **179,** 197–203.
11. Buttery, L. D. K., Springall, D. R., Chester, A. H., Evans, T. J., Standfield, N., Parums, D. V., Yacoub, M. H., and Polak, J. M. (1996) Inducible NO synthase is present within human atherosclerotic lesions and promotes the formation and activity of peroxynitrite. *Lab. Invest.* **75,** 77–85.
12. Higman, D. J., Strachan, A. M. J., Buttery, L. D. K., Hicks, R. C. J., Springall, D. R., Greenhalgh, R. M., and Powell, J. T. (1996) Smoking impairs the activity of endothelial nitric oxide synthase in saphenous vein. *Arterioscler. Thromb. Vasc. Biol.* **16,** 546–552.
13. Evans, T. J., Buttery, L. D. K., Carpenter, A., Springall, D. R., Polak, J. M., and Cohen J. (1996) Inducible nitric oxide synthase is localized within primary granules of cytokine-treated human neutrophils and produces peroxynitrite-mediated nitration of ingested bacteria. *Proc. Natl. Acad. Sci.* **93,** 9553–9558.
14. Coers, W., Timens, W., Kempinga, C., Klok, P., and Moshage, H. (1998) Specificity of antibodies to nitric oxide synthase isoforms in human, guinea pig, rat and mouse tissues. *J. Histochem. Cytochem.* **46,** 1385–1391.
15. Mason, N. A., Springall, D. R., Burke, M., Pollock, J., Mikhail, G., Yacoub, M. H., and Polak, J. M. (1998) High expression of endothelial nitric oxide synthase in plexiform lesions of pulmonary hypertension. *J. Pathol.* **185** 313–318.

14

Anti-Nitrotyrosine Antibodies for Immunohistochemistry

Liliana Viera, Yao Zu Ye, and Joseph S. Beckman

1. Introduction

Nitrotyrosine is an important marker for the formation of peroxynitrite and possibly other reactive nitrogen species derived from nitric oxide in vivo *(1)*. Pathological conditions can substantially increase the production of nitric oxide, yet this molecule itself does not generally yield nitration of tyrosine residues in proteins when added to biological samples *(1,2)*. However nitric oxide reacts at near diffusion-limited rates with superoxide (O_2^-) to form the strong oxidant peroxynitrite ($ONOO^-$) *(3)*. Nitration on the 3-position of tyrosine is a major product of peroxynitrite attack on proteins *(4,5)*. Certainly, small amounts of nitrotyrosine can be produced in vivo by other mechanisms *(6)*, but peroxynitrite is by far the most efficient mechanism for nitrating tyrosine under biologically relevant conditions with natural antioxidants and alternative targets present.

Several functional consequences of nitration have been so far elucidated. Nitration of tyrosine residues disrupts phosphorylation in vitro, which is crucial for signal transduction pathways *(7–10)*. Peroxynitrite modifies low-density lipoprotein (LDL) *(11)* and thrombin *(12)*. It also profoundly disrupts alveolar surfactant function *(13)* and affects assembly of structural proteins like actin *(14,15)* and neurofilament L *(16,17)*.

We have developed specific monoclonal and polyclonal antibodies to recognize 3-nitrotyrosine in proteins *(18,19)*. These antibodies can be used in enzyme-linked immunoabsorbent assays (ELISA), Western blots, and immunohistochemistry of frozen and paraffin-embedded-fixed tissues, both in human and other mammals.

From: *Methods in Molecular Medicine, Vol. 36: Septic Shock*
Edited by: T. J. Evans © Humana Press Inc., Totowa, NJ

Nitric oxide production is easy to detect in rodent inflammatory cells, but more difficult to find in humans. Using these antibodies as well as detection of nitrotyrosine by complementary techniques (HPLC, mass spectrometry, and so on), we and others have provided direct evidence to show that human inflammatory cells produce substantial amounts of nitric oxide-derived oxidants in human disease.

Nitrotyrosine antibodies have proved to be accurate and efficient tools to demonstrate nitration in human tissues with different pathological conditions as well as in animal and cellular models of disease ***(20–39)***.

Several powerful controls using blocking antigens and direct chemical modification of nitrotyrosine offer strong evidence to confirm the specificity of the nitrotyrosine detection by immunohistochemistry and in Western blots. The antibodies reveal the distribution of nitrotyrosine in tissues and the surprising specificity as to which proteins are nitrated in vivo, which cannot be identified by other analytical methods.

1.1. Preparation of the Antibodies

1.1.1. Polyclonal Antibodies

The preparation of the antigen used to raise the antibodies is based upon the nitration of keyhole limpet hemocyanin with 1 m*M* peroxynitrite in the presence of 1 m*M* ferric-EDTA as described previously ***(18,19)***. Peroxynitrite was prepared from quenching the reaction of acidified nitrate and hydrogen peroxide with sodium hydroxide as described previously ***(4)***. Rabbits were injected with peroxynitrite-modified keyhole limpet hemocyanin (Pierce, Rockville, IL). Keyhole limpet hemocyanin (8 mg) in 5 mL of 100 m*M* potassium phosphate (pH 7.4) plus 1 m*M* Fe^{3+} EDTA was mixed rapidly with 2 m*M* peroxynitrite (final concentration). After dialysis overnight, the protein was mixed with an equal volume of Freund's complete adjuvant to a final concentration of 0.5 mg/mL. Rabbits were boosted biweekly with 0.5 mg nitrated keyhole limpet hemocyanin mixed with an equal volume of Freund's incomplete adjuvant. IgG was purified with Gamma Bind G affinity columns from Pharmacia (Uppsala, Sweden) and dialyzed against phosphate-buffered saline overnight.

1.1.2. Monoclonal Antibodies

Mice (BALB/C, Charles River Breeding Laboratories, Wilmington, MA) were immunized with the same peroxynitrite-modified keyhole limpet hemocyanin according to Lieberman ***(40)*** with slight modifications. The primary immunization was 1 mg/mL of nitrated keyhole limpet hemocyanin in phosphate-buffered saline (PBS) emulsified with an equal volume of Freund's complete adjuvant (GIBCO, Grand Island, NY). Each mouse received 0.3 mL of the vaccine delivered subcutaneously in the rear hind foot pads and over the abdominal area. Seven days later the mice were injected in a similar manner with 100 μg nitrated keyhole limpet hemocyanin in PBS alone. Approximately 16 d after the first immunization, the spleen and popliteal lymph nodes were removed, prepared as a cell suspension, and fused with the nonsecreting cell line P3X63-Ag8.653 as described by Kearney ***(41)***. After fusion, cells were seeded into 24-well tissue-culture plates (Corning Costar Corporation, Cambridge, MA) in plating media consisting of RPMI 1640 with 15% fetal calf serum, 5×10^{-5} *M* 2-mercaptoethanol, 100 U/mL penicillin, 100 μg streptomycin and HAT (ICN Pharmaceuticals, Costa Mesa, CA). Fusions were fed on days 7 and 9 by replacing half of the remaining media with fresh HAT media. The last feeding, 48 h prior to screening, consisted of a complete change of medium. On day 14 after fusion, supernatants were screened for the combination of binding to peroxynitrite-modified bovine serum albumin (BSA), the lack of binding to BSA and blockage of binding by 10 m*M* nitrotyrosine by ELISA. Positive wells were cloned by limiting dilution on feeder cells obtained from the peritoneal cavity of pristine-primed mice. Hybridoma cell lines were raised as ascites in pristine-primed BALB/c mice. A suitable colony was used for preparing ascites fluid and the IgG fraction by Protein A chromatography was also purified.

The antibodies are stored in a non-freeze thaw freezer at –80°C for up to several years. Both monoclonal and polyclonal antibodies are commercially available from several sources (Upstate Biologicals, Lake Placid, New York; Cayman Biochemicals, Ann Arbor, MI).

1.2. Specificity of the Antibodies

One concern raised commonly when using nitrotyrosine antibodies is that staining may be diffuse and mistaken as being nonspecific. The reason for this pattern is that the nitrotyrosine epitope is small and is present on many types of proteins. The oxidants that produce nitrotyrosine are able to diffuse over distances of at least one cell diameter. That is why the staining pattern will be more widespread than expected when compared to antibodies directed against more specific proteins. Also, the amount of nitrotyrosine present in tissue samples can vary widely, which necessitates experimentation with different dilutions of the antibody and incubation times to optimize staining.

To determine whether antibodies are specifically recognizing nitrotyrosine, blocking controls are essential. These type of controls prevent specific binding of the nitrotyrosine antibody and will reveal nonspecific interactions that may occur such as binding of the secondary antibodies.

Antibody binding can be blocked by 1 to 10 m*M* concentration of the amino acid 3-nitrotyrosine. These high concentrations are required because the amino acid side chain of the free tyrosine interacts with the aromatic ring and hydroxyl group. Similar concentrations of phosphotyrosine are necessary to inhibit the binding of phosphotyrosine antibodies. A 100-1000 fold lower concentration of the tripeptide glycine-3-nitrotyrosine-alanine, is generally effective at blocking antibody binding ***(18)***. As a rule, the nitrotyrosine antibodies prepared from peroxynitrite-treated keyhole limpet hemocyanin are much more effective at recognizing nitrotyrosine in proteins than the free amino acid nitrotyrosine.

A useful independent control is to treat the tissue sections with a strong reducing agent like sodium hydrosulfite, commonly known as dithionite. This compound is a caustic and highly reactive reductant that is sensitive to both air and moisture. It should eliminate antibody binding because the 3-nitrotyrosine is reduced to 3-aminotyrosine, which is not recognized by the antibody. Prolonged exposure of aminotyrosine to oxygen or mild oxidation will result in the reappearance of antibody binding as aminotyrosine oxidizes readily to nitrotyrosine, although the recovered immunoreactivity will be significantly weaker. The antibody does not cross react with 3,4-dihydrotyrosine, phosphotyrosine, tyramine, 3-methyltyrosine or 3-chlorotyrosine ***(19)***.

Positive controls for nitrotyrosine are generated easily in fixed tissue samples by directly nitrating tyrosine in tissue sections with peroxynitrite or acidified nitrite plus hydrogen peroxide (which generates peroxynitrite *in situ*). Alternatively, the more toxic and volatile tetranitromethane can also be used (*see* **Subheading 6.2.** for further details). Any of these methods should yield a very strong immunoreactivity for nitrotyrosine on any treated slide. This immunoreactivity should be abolished blocking the antibody binding to nitrotyrosine. *In situ* nitration can reveal relative differences in the amounts of tyrosine found in tissues, which could account for some differences in the localization of nitrotyrosine.

2. Materials

2.1. Buffers

1. PBS: 0.1 *M* at pH 7.4 with/without 0.3% Triton-X100 or 0.2% or Tween 20 (we recommend adding a detergent to permeabilize cell membranes in cell preparations).
2. 0.05 *M* Tris-HCl at pH 7.6.
3. 100 m*M* sodium borate buffer at pH 9.0.
4. 50 m*M* sodium borate buffer at pH 8.0.
5. 50 m*M* phosphate buffer at pH 7.4.
6. 100 m*M* acetate buffer at pH 5.0.

2.2. Fixation of Tissue Samples

1. For most tissues the recommended fixative is 4% paraformaldehyde diluted in PBS. For brain and spinal cord we recommend methacarn (chloroform/methanol/acetic acid, 60:30:10) at 4°C overnight *(35)*.
2. Sucrose 15 and 30% solution in PBS.

2.3. Immunoperoxidase

1. 100% methanol at –20°C, plus 0.3% hydrogen peroxide.
2. 10% goat serum: Store the undiluted serum at –20°C in small aliquots. Dilute to 10% in PBS 0.1% at pH 7.4. Filter through a 0.45-µm membrane.
3. Dilution buffer: 5% goat serum + 1% BSA in PBS.
4. DAB solution (5mg/mL): 3-3'-diaminobenzidine tetrahydrochloride (DAB tablets, Sigma). Dissolve one tablet in 20 mL TRIS with stirring. Centrifuge for 10 min at maximum speed. Filter through a 0.45-µm membrane. Just before use mix 10 mL of the solution with 1.5 µL 30% H_2O_2 (0.005% final concentration). A slight enhancement of the reaction can be obtained adding nickel chloride (0.2%) to the final solution: 9 mL DAB + 4 µL 30% H_2O_2 + 1 mL 0.2% nickel chloride.

3. Methods

3.1. Peroxidase-Based Immunohistochemistry Protocol for Paraffin-Embedded Tissues

1. For most tissues fixation can be carried out either by perfusion or immersion in a freshly prepared solution of 4% paraformaldehyde (*see* **Note 1**). The elective method is to fix the tissue by perfusion followed by a postfixation step of 2 h in the same type of fixative. Immersion requires the tissue to be chopped into small pieces and immersed in the fixative preferably for no longer than overnight. The tissue should then be transferred to 70% ethanol and kept in the refrigerator until paraffin embedding. The samples are then dehydrated in a graded ethanol series, embedded in paraffin (Paraplast, 56°C melting point, Oxford Labware, St. Louis, MO) and sectioned at 5–10 µm.
2. To quench endogenous peroxidase activity, incubate rehydrated sections with methanol 100% at –20°C with 0.3% H_2O_2 for 30 min.
3. Wash in PBS for 3× 10 min with gentle stirring.
4. To block nonspecific binding incubate with 10% goat serum for 1 h (*see* **Note 2**).
5. Incubate with the primary antibody (nitrotyrosine monoclonal or polyclonal antibody) (*see* **Note 3**) diluted in dilution buffer, overnight at 4°C in a humid chamber (*see* **Note 4**).
6. Wash in PBS for 3× 10 min with gentle stirring.
7. Incubate with the biotinylated secondary antibody (goat anti-rabbit or anti-mouse IgG) for 1 h. This antibody can be peroxidase-conjugated goat anti-rabbit or anti-mouse IgG (Boehringer-Mannheim, Indianapolis, IN). If the secondary antibody is not biotinylated, a third step is necessary, using a streptavidin horseradish peroxidase (HRP)-conjugated antibody (Pierce, Rockford, IL) (*see* **Note 5**).

8. When using the nonbiotinylated antibody wash with PBS 3 times for 10 min each between the secondary antibody and the streptavidin-HRP-conjugated antibody.
9. To develop the reaction, the slides are incubated in DAB solution (*see* **Note 6**) and 0.005% H_2O_2 in Tris-HCl for several minutes until a brown color develops. The intensity of the reaction can be monitored under a microscope.
10. Rinse in distilled water several times (*see* **Note 7**).
11. Counterstain with hematoxylin, methyl green, or neutral red if desired.
12. The slides are then dehydrated through ascending ethanol series, mounted using a permanent mounting medium and cover slipped.

3.2. Peroxidase-Based Immunohistochemistry Methods on Frozen Sections

1. Tissue is fixed either by perfusion or immersion in paraformaldehyde 4% according to protocol, washed in PBS and cryoprotected in increasing concentrations of sucrose (15–30%) until the tissue sinks.
2. Place the tissue in an embedding cassette containing tissue-freezing medium (Tissue-Tek O.C.T. Compound, Sakura Finetek) and orientate the sample. Immerse the cassette in isopentane precooled in a mix of dry ice and acetone until the tissue is solid and turns white. The samples can be stored at –70°C for long periods of time in air-tight containers with desiccants.
3. In the cryostat, set the temperature according to the characteristics of your sample and allow the tissue to equilibrate with the temperature in the chamber. The thickness of the sections should be between 5 and 10 µm.
4. Collect the sections on electrostatic adherent slides (Superfrost Plus, Fisher, Pittsburgh, PA) and allow them to air dry for at least half an hour.
5. Follow **steps 2–12** in **Subheading 3.1.**

3.3. Protocol for Indirect Immunofluorescence on Frozen or Paraffin-Embedded Tissue

1. For paraffin-embedded sections, deparaffinize the slides as usual from ethanol to PBS. For frozen sections rehydrate the sections in PBS.
2. Incubate the sections in 100% methanol at –20°C with 0.3% H_2O_2 for 30 min.
3. Wash in PBS 2× 5 min with gentle stirring.
4. Incubate with 10% goat serum for 1 h.
5. Incubate with the primary antibody, overnight at 4°C in a humid chamber.
6. Wash in PBS 3× 5 min with gentle stirring.
7. Incubate with the secondary antibody (*see* **Note 8**) for 1 h at room temperature.
8. Wash in PBS 2× 5 min with gentle stirring.
9. Postfix the slides with 4% paraformaldehyde for 5 min (*see* **Note 9**).
10. Wash in PBS 3× 5 min.
11. Mount in an aqueous mounting medium (SlowFade-Light or ProLong Antifade Kit, Molecular Probes, Eugene, OR). If the medium is not permanent, seal around the cover slip with nail polish.

3.4. Controls

3.4.1. Specificity of Binding

1. The primary antibody should be blocked by mixing with either the free aminoacid nitrotyrosine (1–10 m*M*) (*see* **Note 10**) or more efficiently with nitrotyrosine containing peptides (10–100 μ*M*). These must be custom synthesized, or made by adding sequential additions of peroxynitrite to commercially available peptides in a sodium carbonate buffer (50 m*M* at pH 7.0) with vigorous vortexing *(42)*.

   Pre-adsorb the antibodies with any of these blockers at least 3 h before adding to the slide and proceed with the protocol as usual.
2. The nitrotyrosine can be reduced *in situ* to aminotyrosine with sodium hydrosulphite (dithionite) (*see* **Note 11**).

   We recommend dissolving 100 m*M* dithionite in 100-m*M* sodium borate at pH 9, in a sealed degassed vacuttainer tube. The dithionite is prepared within an hour of usage. One milliliter is withdrawn with a tuberculin syringe and the slide is washed three times with about a third of a milliliter used each time.

3.4.2. Positive Controls

1. The simplest method is to mix peroxynitrite and 50 m*M* phosphate buffer at pH 7.4 (1 m*M* final concentration) on top of the rehydrated tissue sample. This will result in nitration of the tissue in a few seconds.
2. Alternatively, add between 100–300 μ*M* tetranitromethane diluted in 50 m*M* sodium borate buffer at pH 8.0, to the slides for 5–15 min, at room temperature, immediately after deparaffinizing. Prepare a stock solution of 100 m*M* tetranitromethane in 95% ethanol and keep it at –20°C (*see* **Note 12**).
3. If peroxynitrite is not readily available, it can be generated *in situ* mixing equal volumes of 2 m*M* sodium nitrite plus 2 m*M* H_2O_2 in 100 m*M* acetate buffer at pH 5.0, and incubate the slides for 5–20 min at room temperature, immediately after deparaffinizing.

4. Notes

1. The antibody has shown to be successful even in autopsy specimens fixed by immersion in neutral-buffered formaldehyde for long periods of time. Short fixation is still preferable.
2. After the incubation in goat serum do not wash the slides, simply wipe it around the section and add the primary antibody.
3. For initial staining of the tissues, we recommend the titration of the antibody using serial dilutions (from 1:50 to 1:1000 of a 1 mg/mL stock in dilution buffer) to produce a suitable signal. The titration must also be performed with every new batch of antibody. A short centrifugation step (10,000*g* for 5 min) is usually recommended before adding the primary and the secondary antibodies to the slides.

   As the monoclonal antibody usually gives weaker immunoreactivity compared to the polyclonal, it is a good practice to carry out the initial experiments with the

polyclonal antibody (except when working with rabbit tissue) and then use the monoclonal as a further verification of specificity.
4. Incubation overnight at 4°C is our standard protocol. We have also tried shorter incubation times at 37°C (from 30 min to 3 h) with good results. In case the expected signal is very weak the incubation can be carried out overnight at room temperature having in mind that this method may increase the nonspecific background.
5. Usually the standard commercial kits based upon avidin-biotin-peroxidase method (DAKO Corporation, Carpinteria, California; Vector Laboratories) give excellent results.
6. DAB is a suspected carcinogen. Follow strict procedures for handling and disposal of the substance.
7. Washing extensively after the developing step helps to get rid of the excess DAB and to prolong the slide preservation in time.
8. The secondary antibody should be titrated in the same experimental conditions, keeping the primary antibody at the chosen working dilution.
9. The postfixation step improves the fluorescence and avoids rapid quenching.
10. It is important to re-adjust the pH of the PBS or TRIS-HCl after adding the free nitrotyrosine.
11. Dithionite is a caustic and highly reactive reductant that is sensitive to both air and moisture. Appropriate care including gloves, must be exercised while handling this compound. The dithionite powder is best aliquoted into small scintillation vials, which are stored in a vacuum desiccator. If the powder is not free-flowing, the vial should be discarded.
12. Tetranitromethane is volatile and toxic. It is necessary to handle it carefully in a fume hood.

References

1. Beckman, J. S. (1996) Oxidative damage and tyrosine nitration from peroxynitrite. *Chem. Res. Toxicol.* **9,** 836–844.
2. Gunther, M. R., Hsi, L. C., Curtis, J. F., Gierse, J. K., Marnett, L. J., Fling, T. E., and Mason, R. P. (1997) Nitric oxide trapping of the tyrosyl radical of prostaglandin H synthase-2 leads to tyrosine iminoxyl radical and nitrotyrosine formation. *J. Biol. Chem.* **272,** 17,086–17,090.
3. Huie, R. E. and Padmaja, S. (1993) The reaction rate of nitric oxide with superoxide. *Free Rad. Res. Commun.* **18,** 195–199.
4. Beckman, J. S., Ischiropoulos, H., Zhu, L., van der Woerd, M., Smith, C., Chen, J., Harrison, J., Martin, J. C., and Tsai, M. (1992) Kinetics of superoxide dismutase- and iron-catalyzed nitration of phenolics by peroxynitrite. *Arch. Biochem. Biophys.* **298,** 438–445.
5. Ischiropoulos, H., Zhu, L., Chen, J., Tsai, H. M., Martin, J. C., Smith, C. D., and Beckman, J. S. (1992) Peroxynitrite-mediated tyrosine nitration catalyzed by superoxide dismutase. *Arch. Biochem. Biophys.* **298,** 431–437.

6. Eiserich, J. P., Cross, C. E., Jones, A. D., Halliwell, B., and van der Vliet, A. (1996) Formation of nitrating and chlorinating species by reaction of nitrite with hypochlorous acid: a novel mechanism for nitric oxide-mediated protein modification. *J. Biol. Chem.* **271,** 19,199–19,208.
7. Martin, B. L., Wu, D., Jakes, S., and Graves, D. J. (1990) Chemical influences on the specificity of tyrosine phosphorylation. *J. Biol. Chem.* **265,** 7108–7111.
8. Berlett, B. S., Friguet, B., Yim, M. B., Chock, P. B., and Stadtman, E. R. (1996) Peroxynitrite-mediated nitration of tyrosine residues in *Escherichia coli* glutamine synthetase mimics adenylalation: Relevance to signal transduction. *Proc. Natl. Acad. Sci. USA* **93,** 1776–1780.
9. Gow, A. J., Duran, D., Malcolm, S., and Ischiropoulos, H. (1996) Effects of peroxynitrite-induced protein modifications on tyrosine phosphorylation and degradation. *FEBS Lett.* **385,** 63–66.
10. Kong, S.-K., Yim, M. B., Stadtman, E. R., and Chock, P. B. (1996) Peroxynitrite disables the tyrosine phosphorylation regulatory mechanism: lymphocyte-specific tyrosine kinase fails to phosphorylate nitrated cdc2(6-20)NH_2 peptide. *Proc. Natl. Acad. Sci. USA* **93,** 3377–3382.
11. Leeuwenburgh, C., Hardy, M. M., Hazen, S. L., Wagner, P., Oh-ishit, S., Steinbrecher, U. P., and Heinecke, J. W. (1997) Reactive nitrogen intermediates promote low density lipoprotein oxidation in human atherosclerotic intima. *J. Biol. Chem.* **272,** 1433–1436.
12. Lundblad, R. L., Noyes, C. M., Featherstone, G. L., Harrison, J. H., and Jenzano, J. W. (1988) The reaction of alpha-thrombin with tetranitromethane. *J. Biol. Chem.* **263,** 3729–3734.
13. Haddad, I. Y., Crow, J. P., Hu, P., Ye, Y. Z., Beckman, J. S., and Matalon, S. (1994) Concurrent generation of nitric oxide and superoxide damages surfactant protein A (SP-A). *Am. J. Phys.* **267,** L242–L249.
14. Chantler, P. D. and Gratzer, W. B. (1975) Effects of specific chemical modification of actin. *Eur. J. Biochem.* **60,** 67–72.
15. Miki, M., Barden, J. A., dos Remedios, C. G., Phillips, L., and Hambly, B. D. (1987) Interaction of phalloidin with chemically modified actin. *Eur. J. Biochem.* **165,** 125–130.
16. Crow, J. P., Strong, M. J., Zhuang, Y., Ye, Y., and Beckman, J. S. (1997) Superoxide dismutase catalyzes nitration of tyrosines by peroxynitrite in the rod and head domains of neurofilament L. *J. Neurochem.* **69,** 1945–1953.
17. Crow, J. P., Sampson, J. B., Zhuang, Y., Thompson, J. A., and Beckman, J. S. (1997) Decreased zinc affinity of amyotrophic lateral sclerosis-associated superoxide dismutase mutants leads to enhanced catalysis of tyrosine nitration by peroxynitrite. *J. Neurochem.* **69,** 1936–1944.
18. Ye, Y. Z., Strong, M., Huang, Z.-Q., and Beckman, J. S. (1996) Antibodies that recognize nitrotyrosine. *Methods Enzymol.* **269,** 201–209.
19. Beckman, J. S., Ye, Y. Z., Anderson, P., Chen, J., Accavetti, M. A., Tarpey, M. M., and White, C. R. (1994) Extensive nitration of protein tyrosines in human

atherosclerosis detected by immunohistochemistry. *Biol. Chem. Hoppe-Seyler* **375,** 81–88.
20. Kooy, N. W., Royall, J. A., Ye, Y. Z., Kelly, D. R., and Beckman, J. S. (1995) Evidence for *in vivo* peroxynitrite production in human acute lung injury. *Am. J. Respir. Crit. Care. Med.* **151,** 1250–1254.
21. Haddad, I., Pataki, G., Hu, P., Galliani, C., Beckman, J. S., and Matalon, S. (1994) Quantitation of nitrotyrosine levels in lung sections of patients and animals with acute lung injury. *J. Clin. Invest* **94,** 2407–2413.
22. MacMillan-Crow, L. A., Crow, J. P., Kerby, J. D., Beckman, J. S., and Thompson, J. A. (1996) Nitration and inactivation of manganese superoxide dismutase in chronic rejection of human renal allografts. *Proc. Natl. Acad. Sci. USA* **93,** 11,853–11,858.
23. Kooy, N. W., Lewis, S. J., Royall, J. A., Ye, Y. Z., Kelly, D. R., and Beckman, J. S. (1997) Extensive tyrosine nitration in human myocardial inflammation: evidence for the presence of peroxynitrite. *Crit. Care Med.* **25,** 812–819.
24. Ishiyama, S., Hiroe, M., Nishikawa, T., Abe, S., Shimojo, T., Ito, H., Ozasa, S., Yamakawa, K., Matsuzaki, M., Mohammed, M. U., Nakazawa, H., Kasajima, T., and Marumo, F. (1997) Nitric oxide contributes to the progression of myocardial damage in experimental autoimmune myocarditis in rats. *Circulation* **95,** 489–496.
25. Myatt, L., Rosenfield, R. B., Eis, A. L. W., Brockman, D. E., Greer, I., and Lyall, F. (1996) Nitrotyrosine residues in placenta. Evidence of peroxynitrite formation and action. *Hypertension* **28,** 488–493.
26. Miller, M. J. S., Thompson, J. H., Zhang, X. J., Sadowska-Krowicka, H., Kakkis, J. L., Munshi, U. K., Rossi, J. L., Eloby-Childress, S., Beckman, J. S., Ye, Y. Z., Roddi, C. P., Manning, P. T., Currie, M. G., and Clark, D. A. (1995) Role of inducible nitric oxide synthase expression and peroxynitrite formation in guinea pig ileitis. *Gastroenterology* **109,** 1475–1483.
27. Kaur, H. and Halliwell, B. (1994) Evidence for nitric oxide-mediated oxidative damage in chronic inflammation. Nitrotyrosine in serum and synovial fluid from rheumatoid patients. *FEBS Lett.* **350,** 9–12.
28. Halliwell, B. (1995) Oxygen radicals, nitric oxide and human inflammatory joint disease. *Ann. Rheum. Dis.* **54,** 505–510.
29. Hantraye, P., Brouillet, E., Ferrante, R., Palfi, S., Bolan, R., Matthews, R. T., and Beal, M. F. (1996) Inhibition of neuronal nitric oxide synthase prevents MPTP-induced parkinsonism in baboons. *Nat. Med.* **2,** 1017–1021.
30. Ischiropoulos, H., Beers, M. F., Ohnishi, S. T., Fisher, D., Garner, S. E., and Thom, S. R. (1996) Nitric oxide production and perivascular tyrosine nitration in brain following carbon monoxide poisoning in the rat. *J. Clin. Invest.* **97,** 2260–2267.
31. Gow, A., Duran, D., Thom, S. R., and Ischiropoulos, H. (1996) Carbon dioxide enhancement of peroxynitrite-mediated protein tyrosine nitration. *Arch. Biochem. Biophys.* **331,** 42–48.

32. Beal, M. F., Ferrante, R. J., Henshaw, R., Matthews, R. T., Chan, P. H., Kowall, N. W., Epstein, C. J., and Schulz, J. B. (1995) 3-nitropropionic acid neurotoxicity is attenuated in copper/zinc superoxide dismutase transgenic mice. *J. Neurochem.* **65,** 919–922.
33. Good, P. F., Werner, P., Hsu, A., Olanow, C. W., and Perl, D. P. (1996) Evidence for neuronal oxidative damage in Alzheimer's disease. *Am. J. Pathol.* **149,** 21–28.
34. Basarga, O., Michaels, F. H., Zheng, Y. M., Borboski, L. E., Spitsin, S. V., Fu, Z. F., Tawadros, R., and Koprowski, H. (1995) Activation of the inducible form of nitric oxide synthase in the brains of patients with multiple sclerosis. *Proc. Natl. Acad. Sci. USA* **92,** 12,041–12,045.
35. Smith, M. A., Harris, P. L., Sayre, L. M., Beckman, J. S., and Perry, G. (1997) Widespread peroxynitrite-mediated damage in Alzheimer disease. *J. Neurosci.* **17,** 2653–2657.
36. Abe, K., Pan, L.-H., Watanabe, M., Kato, T., and Itoyama, Y. (1995) Induction of nitrotyrosine-like immunoreactivity in the lower motor neuron of amyotrophic lateral sclerosis. *Neurosci. Lett.* **199,** 152–154.
37. Ferrante, R. J., Shinobu, L. A., Schulz, J. B., Matthews, R. T., Thomas, C. E., Kowall, N. W., Gurney, M. E., and Beal, M. F. (1997) Increased 3-nitrotyrosine and oxidative damage in mice with a human copper/zinc superoxide dismutase mutation. *Ann. Neurol.* **42,** 326–334.
38. Beal, M. F., Ferrante, R. J., Browne, S. E., Matthews, R. T., Kowall, N. W., and Brown, R. H., Jr. (1997) Increased 3-nitrotyrosine in both sporadic and familial amyotrophic lateral sclerosis. *Ann. Neurol.* **42,** 646–654.
39. Estévez, A. G., Spear, N., Manuel, S. M., Radi, R., Henderson, C. E., Barbeito, L., and Beckman, J. S. (1998) Nitric oxide and superoxide contribute to motor neuron apoptosis induced by trophic factor deprivation. *J. Neurosci.* **18,** 923–931.
40. Lieberman, R., Potter, M., Humphrey, W., Mushinski, E. B., and Vrana, M. (1975) Multiple individual and cross-specific idiotypes on 13 levan-binding myeloma proteins of BALB/C mice. *J. Exp. Med.* **142,** 106–119.
41. Kearney, J. F. (1984) Hybridomas and monoclonal antibodies, in *Fundamental Immunology*. (Paul, W. E., ed.) Raven Press, New York, pp. 751–766.
42. Sampson, J. B., Rosen, H., and Beckman, J. S. (1996) Peroxynitrite dependent tyrosine nitration catalyzed by superoxide dismutase, myeloperoxidase and horse-radish peroxidase. *Methods Enzymol.* **269,** 210–218.

15

Detection of Peroxynitrite in Biological Fluids

Stuart Malcolm, Raymond Foust III,
Caryn Hertkorn, and Harry Ischiropoulos

1. Introduction

Peroxynitrite ($ONOO^-$) is both an oxidant and a nitrating agent *(1–3)*. However, unlike other strong oxidants, peroxynitrite reacts selectively with biological targets. This selectivity is derived in part from the different second-order rate constants (vary from 10^3–10^6 M^{-1} s^{-1}) by which $ONOO^-$ reacts with biological targets. Competing for peroxynitrite-mediated oxidation of biological targets are two pathways. One is the protonation to form peroxynitrous acid and the second is the reaction with CO_2. Peroxynitrous acid is also an oxidant but it readily isomerizes to nitrate. The reaction with CO_2 results in the formation of the $ONO(O)CO_2^-$ adduct that is a more potent nitrating agent but a weaker oxidant than peroxynitrite. Peroxynitrite can diffuse through biological membranes before it encounters and reacts with biological targets. This observation implies that diffusion can effectively compete with the isomerization to nitrate or the reaction with CO_2.

The detection of peroxynitrite for the most part relies upon the detection of 3-nitrotyrosine because this is the major modification derived from the reaction of peroxynitrite with proteins. In addition to the detection of nitrotyrosine, the use of fluorescence probes may be also suitable for detecting peroxynitrite in vivo *(4–6)*. This relies on the oxidation of the nonfluorescent dye dihydrorhodamine 123 to the fluorescent derivative rhodamine 123. However, both methodologies require the judicious use of inhibitors and a number of controls. For example, nitration can be derived from a number of reactive pathways all dependent on the formation of nitric oxide. Only the peroxynitrite pathway will be dependent on superoxide but not on hydrogen peroxide. Therefore, scavenging of superoxide and hydrogen peroxide independently, together

From: *Methods in Molecular Medicine, Vol. 36: Septic Shock*
Edited by: T. J. Evans © Humana Press Inc., Totowa, NJ

with the quantification of nitrotyrosine will measure the fraction of the peroxynitrite that reacted with protein. Similarly the same controls can be used for the detection of peroxynitrite by fluorescent compounds. Methodologies for the use of fluorescence compounds as well as for quantifying nitrotyrosine are described below.

2. Materials

2.1. Dihydrorhodamine 123 (DHR) for the Detection of Peroxynitrite in Biological Fluids

1. Stock solution of DHR and rhodamine (RH): Stock solutions of dihydrorhodamine 123 and rhodamine 123 (Molecular Probes, Eugene OR) are prepared by dissolving solid DHR in dimethylformamide in a light protected vial vessel that is purged with nitrogen. Nitrogen-purged aliquots are stored at –20°C. Under these storage conditions stock solutions are stable for several months. Working solutions of DHR in the desired buffer should be made fresh every day, tested for oxidation of DHR and kept on ice in light protected vessels. The concentration of the oxidized product RH in working solutions of DHR should be evaluated either spectrophotometrically or fluorometrically.
2. Heparinized tubes.
3. Fluorometer and/or spectrophotometer.
4. Peroxynitrite (Calbiochem and Alexis, San Diego, CA).

2.2. Generation of Protein Nitrotyrosine Standards

1. Fatty-acid-free bovine serum albumin (BSA).
2. Reaction buffer: 100 m*M* sodium phosphate buffer at pH 7.4, 50 m*M* sodium bicarbonate, freshly prepared.
3. Peroxynitrite (Calbiochem and Alexis).

2.3. Solid-Phase ELISA for the Quantification of Protein 3-Nitrotyrosine

1. Bio-dot microfiltration unit (96-well) (Bio-Rad, Hercules, CA).
2. Nitrocellulose.
3. Diazotized paper (Schleicher and Schull, Keene, NH).
4. Tris-buffered saline at pH 7.0: 500 m*M* NaCl, 20 m*M* Tris-HCl at pH 7.0.
5. Wash buffer (TTBS): 0.05% Tween in TBS.
6. Block buffer: 10% dried milk in TBS.
7. Anti-nitrotyrosine antibody: 2 µg/mL polyclonal or monoclonal (Calbiochem, Alexis, and Upstate Biotechnology, Lake Placid, NY) in 0.5% milk in TTBS.
8. Anti-rabbit or anti-mouse IgG-alkaline phosphatase linked antibody at 1:10,000 (Amersham Life Sciences, Arlington Heights, IL).
9. Chemifluorescent alkaline phosphatase substrate (Amersham Life Sciences).

3. Methods

3.1. Detection of Peroxynitrite in Blood from Experimental Animals

1. Previously it was determined that a bolus iv injection of 0.3 mL (2 μmole/kg) DHR solution to rats had no adverse effects and was sufficient for the detection of peroxynitrite formation after induction of in early stages of endotoxemia, hemorrhagic shock and ischemia-reperfusion injury *(7)*.
2. Prepare the injected solution by diluting a stock DHR solution with saline. The injection of DHR should be performed within 20 min before the collection of blood and before the induction of stress to the animals. It is important to allow sufficient time for oxidation to take place but not to surpass 20 min, because rhodamine will enter the cells and be lost from the circulation. To optimize detection, the oxidation of DHR in circulation should be monitored over time following the induction of stress.
3. Collect blood into heparinized tubes, and separate plasma from the red blood cells by a low 10,000*g* centrifugation for 15 min (*see* **Note 1**).
4. Oxidation of DHR to rhodamine can be measured either by the fluorescence properties of rhodamine or its characteristic absorbance. Prepare a standard curve of rhodamine 123 (1–100 n*M*) in biological fluids obtained from untreated animals. Fluorescent emission of rhodamine is measured in a fluorometer. Excitation of rhodamine solutions at 500 nm results in emission at 526 nm. The slit width should be kept as small as possible, typically 2.5–5.0 nm, because both the excitation and emission peaks are narrow. Background plasma fluorescence from control animals, without injection of DHR, should be subtracted from all samples. Compare unknowns with the standard curve to determine the concentration of rhodamine in the sample (*see* **Note 2**).
5. Alternatively measure the concentration of rhodamine produced by measuring its absorbance. Measurement of DHR oxidation spectrophotometrically is sufficiently sensitive for in vitro assay of submicromolar concentrations of $ONOO^-$ and has the advantage of avoiding the nonlinearity seen with fluorescence emission. The peak absorption of rhodamine is at 500 nm and the extinction coefficient at 500 nm is 78,000 M^{-1} cm^{-1}. Prepare a standard curve of rhodamine as described above and calculate the concentrations in samples by comparison to the standard curve. Background plasma absorbance from control animals, without injection of DHR, should be subtracted from all samples.
6. To ascertain that the concentration of DHR delivered to the animals is sufficient, an aliquot of plasma containing DHR and rhodamine should then be oxidized by addition of 100 μ*M* peroxynitrite or 100 μ*M* H_2O_2 plus 10 μM horseradish peroxidase (HRP). Additional increase in fluorescence or absorbance will indicate that there is sufficient DHR loaded into circulation.
7. It is important to perform suitable controls to demonstrate that the oxidation of DHR is caused by peroxynitrite (*see* **Note 3**).

3.2. Solid-Phase ELISA for the Quantification of Protein 3-Nitrotyrosine

This process consists of three steps: immobilization of proteins onto nitrocellulose, reacting with and developing the blot to visualize binding, and quantification of 3-nitrotyrosine by using standards.

3.2.1. Generating Nitrotyrosine BSA Standards

1. Generate the nitrotyrosine in BSA or other proteins, by reaction with peroxynitrite. Dissolve 4 mg/mL fatty acid free-BSA in 100 m*M* phosphate buffer at pH 7.4 that contains freshly prepared 50 m*M* bicarbonate. Add peroxynitrite to a final concentration of 1–2 m*M*. The protein solution should turn yellow, an indication of the formation of 3-nitrotyrosine.
2. An appropriate control can be generated by the reverse order of addition as follows. Add the same concentration of peroxynitrite to the buffer and allow to decompose for 5 min. Then add the protein.
3. To measure the concentration of nitrotyrosine make an aliquot of the reacted protein alkaline with the addition of 1/10 volume 10 *M* NaOH and a second aliquot of the reacted protein acidic by the addition of 1/10 volume 12 *N* HCl. Both protein solutions are then scanned from 300 to 500 nm. The alkaline solution should absorb maximally at 430 nm. To calculate the concentration of 3-nitrotyrosine, the absorbance of protein solution at acidic pH is subtracted form the absorbance at 430 nm at alkaline pH and the resulting absorbance is divided by 4400 M^{-1} cm^{-1} (extinction coefficient of 3-nitrotyrosine at 430 nm). Chemically nitrated proteins are stable indefinitely since the nitro group is not readily oxidized or reduced. However, nitrotyrosine can be chemically reduced to colorless aminotyrosine by dithionite.

3.2.2. Immobilization of Proteins onto Nitrocellulose

The samples of biological fluids are bound to nitrocellulose filters using a 96-well Bio-Dot microfiltration unit (Bio-Rad). To facilitate loading, samples can be placed into a 96-well plate and then transferred to nitrocellulose under vacuum using a multichannel pipet (*see* **Notes 4** and **5**).

1. Dilute the samples in Tris-buffered saline (TBS) at pH 7.0. We found that this buffer allows for best binding of proteins.
2. Load eight different concentrations of the standard (peroxynitrite-treated BSA) in duplicate. The range of standards loaded is usually 15–0.75 ng of 3-nitrotyrosine in 2.2–0.1 µg protein per spot.

3.2.3. Blot Processing and Development

1. After the samples have been immobilized onto the blot, remove the nitrocellulose and wash twice in wash buffer for 10 min.

2. Incubate the blot overnight with blocking buffer at 4°C. After blocking of the nonspecific binding sites, wash the nitrocellulose three times with wash buffer for 10 min.
3. Incubate the blot for 2 h at room temperature with the anti-nitrotyrosine antibody solution that has been preconjugated with a secondary antibody (*see* **Note 6**). The preconjugation is performed as follows; 2 μg/mL anti-nitrotyrosine antibody (affinity-purified polyclonal or monoclonal) diluted in 0.5% milk in TTBS is incubated overnight at 4°C with an anti-rabbit (polyclonal detection) or anti-mouse (monoclonal detection) IgG-alkaline phosphatase-linked antibody (1 : 10,000, Amersham Life Science).
4. Wash the blot twice in wash buffer, then three times with TBS and dry.
5. Develop the blot directly onto a fluorescence scanning imager (STORM 840, Molecular Dynamics) using Amersham's chemifluorescence substrate.

3.2.4. Quantification of 3-Nitrotyrosine

1. Plot the net counts of fluorescence measured in each sample (corrected for background from a sample blank) on a semilogarithmic plot. Use for a standard blank fatty-acid-free BSA that was not reacted with peroxynitrite or BSA reacted with the reverse order of addition with peroxynitrite. For sample blank two conditions can be used: biological fluid from control animal and samples reduced with dithionite (*see* **Note 7**).
2. Determine the concentration of nitrotyrosine in each sample from the linear portion of the sigmoidal curve from the semilog plot of net counts vs antigen concentration of the nitrated BSA standard. Once the concentration of nitrotyrosine per protein spot is determined it can be plotted against the protein concentration per spot. The slope of the line from the linear regression analysis of this plot represents the concentration of nitrotyrosine per mg of protein.

4. Notes

1. Avoid storing biological fluids with azide. Oxidation of azide by heme proteins results in artifactual formation of nitrotyrosine. If possible, we recommend measuring the quantity of protein 3-nitrotyrosine in fresh biological material. Biological fluids should be stored in the presence of protease inhibitors (0.5 *M* PMSF, 1 μg/mL leupeptin, 10 μ*M* lactacystin, and 1 μg/mL aprotinin) at –80°C.
2. A number of constituents in biological fluids may interfere with either the detection of RH. Both the absorbance and fluorescence of serial dilutions of RH in biological fluids should be monitored to establish a linear response of absorbance or fluorescence with concentration.
3. Controls for the detection of peroxynitrite: Peroxynitrite but not its precursors, superoxide and nitric oxide oxidize DHR to rhodamine (*4–7*). Potential oxidants capable of oxidation of DHR include H_2O_2/peroxidase, H_2O_2/metal, and hypochlorous acid (HOCl). Identification of the reactive species responsible for the oxidation can be achieved with an experimental approach that exploits the

differential stability of reactive species and judicious use of inhibitors and oxidant scavengers.

a. The yield should decrease following inhibition of nitric oxide synthesis. Inhibition of nitric oxide synthesis may have unexpected results as the resulting unreacted superoxide may lead to hydrogen peroxide formation and of DHR via a peroxidase- or metal-dependent pathway. However, if superoxide encounters other targets and thus will not form either peroxynitrite or hydrogen peroxide, the yield of oxidized indicator is not expected to change. Use of NOS inhibitors however, requires a prior determination that ·NO is in fact being produced (e.g., nitrite, nitrate measurement), and that ·NO production is indeed inhibited by the NOS inhibitor(s).
b. The yield should decrease by dismutation of superoxide by the addition of sufficient concentration of superoxide dismutase (SOD). SOD mimetics (Calbiochem), small and cell permeable superoxide scavengers can be also used. However, these SOD mimetics must be tested for catalase activity and peroxynitrite scavenging.
c. The yield will not be affected by addition of catalase. The catalase solutions must be checked for contaminating peroxidase activity. The peroxidase activity in the presence of H_2O_2 will increase the yield of DHR oxidation to rhodamine.
d. Hypochlorous acid is generated by the action of myeloperoxidase and thus mostly associated with neutrophils. Neutrophil involvement can be excluded by the use of neutropenic animals. Independent evidence for peroxynitrite formation could be obtained by concomitant immunological detection of nitrotyrosine-containing proteins and excluding neutrophils.

4. Samples should not contain any detergents because they interfere with binding. Binding of protein also decreases by the inclusion of 12 *M* guanadine-HCL or 8 m*M* urea, or boiling. Essentially the proteins in biological fluid should be only diluted in TBS prior to loading. The most important aspect of immobilizing proteins is the amount of protein loaded. The binding capacity of nitrocellulose is 100 μg/cm^2. The best results are obtained by loading samples in the range of 1–50 μg/spot. However, the optimal ratio of protein to amount of antigen present in biological samples must be determined experimentally.
5. Diazotized paper (Schleicher & Schuell), an alternative to nitrocellulose paper, allows the binding of up to 1 mg of protein per spot. However, we have experienced problems with leakage and lateral diffusion of proteins around the spot using the diazotized paper. The diffusion of fluid resulted in loss of sharpness and cross contamination of spots. Moreover, overloading of protein results in masking of antigenic epitopes and thus a decrease in antibody binding.
6. Preconjugation of the primary and secondary antibody improved specificity and reduced background. Blocking with 10% milk also added in the reduction of background.
7. Reduction of nitrotyrosine to aminotyrosine with dithionite: 3-Nitrotyrosine can be reduced to aminotyrosine with dithionite. The dithionite solution can be pre-

pared placing the dithionite crystals in a plastic tube that contains buffer (pH 9.0) and had been purged with nitrogen. Add a small volume of freshly prepared dithionite solution to a final concentration of 0.1 *M* to reduce 3-nitrotyrosine to aminotyrosine. Dithionite-reduced samples should be prepared fresh because of the time dependent autooxidation of aminotyrosine back to nitrotyrosine.

References

1. Beckman, J. S. and Koppenol, W. H. (1996) Nitric oxide, superoxide and peroxynitrite: the good the bad and the ugly. *Am. J. Physiol.* **217,** C1424–1437.
2. Radi, R. (1996) Reactions of nitric oxide with metalloproteins. *Chem. Res. Tox.* **9,** 828–835.
3. Ischiropoulos, H. (1998) Biological tyrosine nitration: a pathophysiological function of nitric oxide and reactive oxygen species. *Arch. Biochem. Biophys.* **356,** 1–11.
4. Kooy, N. W, Ischiropoulos, H., Beckman, J. S., and Royall, J. A. (1994) Peroxynitrite-mediated oxidation of dihydrorhodamine 123. *Free Rad. Biol. Med.* **16,** 149–156.
5. Royall, J. A. and Ischiropoulos, H. (1993) Evaluation of 2',7' dichlorofluorescin and dihydrorhodamine 123 as fluorescent probes for intracellular hydrogen peroxide in culture endothelial cells. *Arch. Biochem. Biophys.* **302,** 348–355.
6. Crow, J. (1997) Dichlorodihydrofluorescein and dihydrorhodamine 123 are sensitive indicators of peroxynitrite in vitro: implications for intracellular measurement of reactive nitrogen and oxygen species. *Nitric Oxide: Biol. and Chem.* **1,** 146–157.
7. Szabo, C., Salzman, A. L., and Ischiropoulos, H. (1995) Peroxynitrite-mediated oxidation of dihydrorhodamine 123 occurs in early stages of endotoxic and hemorrhagic shock and ischemia-reperfusion injury. *FEBS Lett.* **372,** 229–232.

5

Cell Culture Techniques

16

Myocardial Cells in Culture to Study Effects of Cytokines

Kevin A. Krown and Roger A. Sabbadini

1. Introduction

This chapter describes the use of cultured adult primary ventricular cardiomyocytes as a model for cytokine-mediated septic shock. We describe the methods for preparation of adult cardiomyocytes from adult rat hearts for culture. We also describe techniques designed to assess the responses of single cells in culture to the proinflammatory cytokine, TNFα. We describe in detail the use of this cell-culture model to evaluate TNFα-induced voltage-dependent calcium fluxes as well as the expression of genes that mediate the TNFα response. Because cardiomyocyte cultures may be heterogeneous and may include a variety of other cell types (fibroblasts, endothelial cells, etc.), it is sometimes prudent to examine cells on a single-cell basis to ensure that the physiological or molecular process of interest is characteristic of the cardiomyocyte. The issue of heterogeneity is particularly problematic when attempting to examine the expression of low-abundant transcripts using reverse transcriptase-polymerase chain reaction (RT-PCR). In our laboratory, we have employed single-cell RT-PCR as a tool for examining the expression of genes coding for TNFα receptors in isolated myocardial cells. The procedures can be applied to acutely isolated cells as well as those cultured and subjected to chronic treatments. By using these techniques, it is possible to implicate molecular mechanisms/pathways in the physiological responsiveness of the adult ventricular cardiomyocyte to cytokines. Single-cell analysis overcomes the limitations of studies utilizing cardiac tissue. Primary adult cardiomyocytes in culture are also useful in examining other responses of single cells in culture to cytokines or other physiologically relevant molecules. Some of these responses include:

From: *Methods in Molecular Medicine, Vol. 36: Septic Shock*
Edited by: T. J. Evans © Humana Press Inc., Totowa, NJ

1. Dual emission photometry system for measuring calcium transients *(1–3)*.
2. Single-cell RT-PCR to measure gene expression *(2,4)*.
3. Single-cell microgel electrophoresis and the terminal deoxy nucleotide UTP nick end labeling (TUNEL) assay to demonstrate apoptosis *(1,5)*.
4. Patch clamping in whole-cell configuration to identify membrane ion currents *(2,3)*.
5. The use of edge detectors *(6)* to study the contractile function of single cardiomyocytes.
6. The use of reporter plasmids to identify transfected single cells and to study the responses of the transfected cell to physiologically relevant ligands or activators *(7)*.

It is also possible to use primary cultures of cardiomyocytes to study the behavior of large populations of cells. For example, we recently used enzyme-linked immunosorbent assays (ELISA) to quantify cytokine processing and release/secretion from high density platings of cardiomyocytes *(1)*.

In this chapter we focus on the preparation of adult rat primary cardiomyocytes for culture and the subsequent measurement of calcium transients and gene expression. This methodology affords researchers the opportunity to study cardiomyocyte responsiveness to TNFα and other cytokines at the single-cell level.

2. Materials

2.1. Adult Primary Cardiomyocyte Isolation

1. Solution A: KCl (6 m*M*), NaH_2PO_4 (1 m*M*), Na_2HPO_4 (0.2 m*M*), NaCl (128 m*M*), $MgCl_2$ (1.4 m*M*), sodium pyruvate (2 m*M*), sodium HEPES (10 m*M*), dextrose (5.5 m*M*). Dissolve in distilled water. Adjust pH to 7.4.
2. Solution B: Solution A (50 mL), bovine serum albumin (0.35%), 2,3 butanedione oxime (15 m*M*), $CaCl_2$ (0.1 m*M*), depyrogenated collagenase type A (50 mg, 0.31 U/mg, 15.5 U).
3. Solution D: Solution A (30 mL), $CaCl_2$ (0.1 m*M*), bovine serum albumin (1.0%).
4. Solution E: Solution A (250 mL), $CaCl_2$ (1 m*M*).
5. Heparin: 120 mg/mL in 0.9% sodium chloride.
6. Urethane: 1 g/ml in 0.9% sodium chloride.
7. Perfusion apparatus: Langendorff three-chamber perfusion system and peristaltic pump.

2.2. Adult Primary Cardiomyocyte Culture

1. Adult ventricular myocytes are prepared from 200–350 g Sprague-Dawley male rats *(1–3)*.
2. Dulbecco's modified Eagles medium-F12(DMEM/F12) (GIBCO, Grand Island, NY).
3. Mouse laminin (Collaborative Biomedical Products, Bedford, MA).

4. Media supplements: Fetal calf serum (FCS), bovine serum albumin (BSA), kanamycin (0.1 mg/mL), glutamine (0.6 mg/mL), ampicillin (0.05 mg/mL), pyruvate (0.25 mg/mL), and amphotericin B (0.04 mg/mL). There should be two preparations of media: serum-free (DMEM/F12-) used for rinsing cells and either FCS(10%) or BSA (0.1%) for culture purposes (DMEM/F12+).
5. Glass cover slips (31-mm round) are rinsed in 70% ethanol and exposed to ultraviolet light for at least 12 h in a laminar flowhood. Mouse laminin (250–500 µL) (25–50 µg/mL) is placed on the cover slips and allowed to air dry thoroughly in the laminar flow hood for at least 1–2 h.
6. Culture plates: tissue-culture grade (35-mm, 6-well).
7. Culture-room incubator: water-jacketed; maintained at 37°C. Gas mixture consists of 95% air and 5% carbon dioxide.

2.3. Calcium Transients

1. Tyrode's solution: NaCl (140 m*M*), KCl (5.4 m*M*), $MgCl_2$ (0.5 m*M*), $CaCl_2$ (1 m*M*), HEPES (10 m*M*), and Na_2HPO_4 (0.25 m*M*) at pH 7.3.
2. Intracellular Ca^{2+} indicator: 3 µ*M* indo-1 (Molecular Probes, Eugene, OR) in Tyrodes with 1 m*M* Ca^{2+}.
3. Dual-emission photometry system (Photon Technologies, Monmouth Junction, NJ) interfaced to a Nikon Diaphot microscope.
4. Electrical stimulator (Grass Instruments, West Warwick, RI).
5. Platinum electrodes.
6. Tissue chamber including Teflon chamber, aluminum base, polysulfone plastic shelf, plastic lid, magnetic ring, and four securing screws (Biophysica Technologies, Sparks, MD).

2.4. Single-Cell RT-PCR

1. Microcentrifuge tubes (sterilized).
2. First-strand cDNA synthesis buffer: dNTPs (1 m*M*), random hexanucleotide primers (10 µ*M*), RNAsin (50 U), 2.5× reverse-transcriptase buffer (Boehringer Mannheim, Indianapolis, IN) and AMV reverse transcriptase.
3. RT-PCR buffer: specific primers (100 pmol), dNTPs (200 µ*M* of each dATP, dCTP, dTTP, dGTP), $MgCl_2$ (2 m*M*) and Taq polymerase (1.25 U).
4. Programmable thermal cycler.
5. Micropipet puller.
6. Glass micropipets.

2.5. Miscellaneous Agents

1. Lipopolysaccharide (*Escherichia coli* 055:B5; Difco, Detroit, MI).
2. Rat TNFα (Genzyme Diagnostics, Cambridge, MA).
3. TNF protease inhibitor (TAPI) and rh TNFRII:Fc receptor fragment (dimeric human TNF receptor p80/IgG1 Fc fusion protein; Immunex Corporation, Seattle, WA).

3. Methods

3.1. Adult Primary Cardiomyocyte Isolation (1,2,5)

1. Inject rats intraperitoneally with heparin (0.5 mL) at least 20 min prior to sacrifice (*see* **Note 1**).
3. Inject rats intraperitoneally with urethane (0.7 mL). Wait until animal is completely anesthetized (examine for withdrawal reflex).
4. Following cervical dislocation, quickly place rat on its back. Pinch skin and muscle under the abdomen. Using blunt scissors, cut through the skin and muscle upward and across the chest passing through the sternum.
5. Cut connective tissue between sternum and heart.
6. Lift up heart and locate aorta. Cut 1.5 cm above the atria (behind aortic branches) which will facilitate insertion of cannula of the perfusion apparatus into the aorta.
7. Immediately after the heart is excised, rinse it in 20–25 mL of cold pre-oxygenated solution A (or 0.9% sodium chloride). Using two iris tweezers, grab the aorta and insert it on the cannula of the perfusion apparatus and tie it down with surgical suture (*see* **Note 2**). Start the peristaltic pump, perfusing with solution A (37°C) for 5 min (flow rate 5 mL/min).
8. Switch stopcock on perfusion apparatus and perfuse with solution B (flow rate 1–2 drops/s). Digest the heart for at least 20–30 min (*see* **Notes 3** and **4**).
9. Cut heart from the cannula, just below the atria and transfer heart to a Petri dish containing 20 mL of solution B (35°C, obtained from the jacketed perfusion well). Swirl.
10. Cut ventricles into smaller chunks with scissors.
11. Transfer tissue with the 20 mL of solution B to a 100-mL beaker.
12. Separate cells by gently resuspending (triturating) several times with a 5-mL pipet (*see* **Note 5**).
13. Examine an aliquot of cell suspension under the microscope. If adequately digested, cells should have a rod-shaped morphology (absence of frayed edges) with sarcomeres clearly visible (*see* **Note 6**).
14. If digestion is adequate, add 20 mL of solution D to the beaker (total volume of the beaker should be 40 mL). Solution D quenches the digestion.
15. Transfer the entire volume in the beaker to a 50-mL plastic centrifuge tube by filtering the digested suspension through a 292 μM mesh. The mesh prevents large chunks of undigested muscle from transferred to the cell suspension.
16. Allow cells to settle for at least 15 min.
17. Replace half the volume of the solution B and D mixture with solution E. This will dilute out the collagenase and slowly increase the Ca^{2+} concentration. A gradual increase in Ca^{2+} concentration is essential to ensure a Ca^{2+}-tolerant cell population.
18. Allow cells to settle again and replace half the volume with solution E.
19. Allow cells to settle again and then remove as much of the supernatant as possible while leaving the pellet undisturbed.

20. Add desired volume of solution E. Ca^{2+} concentration is now 1 m*M*.
21. Cells are now ready for culturing.

3.2. Adult Primary Cardiomyocyte Culture

1. A 500–1000 µL suspension of freshly dissociated ventricular cardiomyocytes is plated on laminin-coated (25–50 µg/mL) cover slips that have been placed inside 35-mm wells of a six-well culture plate. Allow cells to adhere to the coverslip for at least 30 min in the laminar flow hood.
2. One to two milliliters of supplemented DMEM/F12 media is added to the wells containing the cardiomyocytes. Agitating plates mildly with the media will cause nonviable cells to detach. Remove media and replace with fresh media and agitate/rinse again. Replace media. Cells are cultured in DMEM/F12+ in the presence or absence of the test agent and placed in the incubator.

3.3. Calcium Transients (1–3)

1. Adult cardiomyocytes plated on 31-mm glass cover slips are loaded for 15–30 min with 500 µL of indo-1 solution and carefully transferred with forceps into the tissue chamber (*see* **Note 7**).
2. After washing and incubation in 500 µL of Tyrodes with 1 m*M* Ca^{2+}, platinum electrodes are placed into the Tyrode's solution and the cells are electrically paced to contract at 0.3 Hz and observed with a light microscope to ensure that the chosen single cardiomyocyte is viable. Viability at this level is determined by contractile response to electrical stimulation and myocyte morphology (*see* **Note 6**).
3. Prior to the addition of the desired test agent, cells are washed and placed in 500 µL of Tyrodes or Ca^{2+}-free Tyrodes (so that the resulting Ca^{2+} transients are a consequence of changes in the intracellular calcium stores during contractions, and not affected by extracellular Ca^{2+}). Test agents are directly added to the chamber at desired time intervals.
4. Fluorescence measurements are performed with a dual emission photometry (Photon Technologies) interfaced to a Nikon Diaphot microscope and a 100X oil immersion objective. For the indo-1 measurements, excitation is at 355 nm. Fluorescence emission is split and monitored simultaneously at 405 and 485 nm at a data collection rate of 20–60 Hz and presented as the 405/485 emission ratio for the indicated period of time. **Figure 1** shows a calcium transient profile from an electrically paced cardiomyocyte. Note that in response to the cytokine, TNFα, there is a depression of transient amplitude which is consistent with the negative inotropic actions of TNFα.

3.4. Single-Cell RT-PCR

The following procedure *(2,4)* is utilized to measure expression of myocardial-specific genes. The procedure involves RNA harvested from individual adult cardiac myocytes as described above *(1,2,5)*.

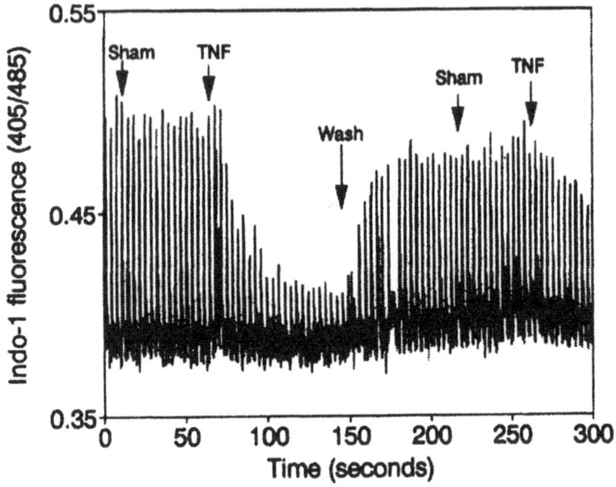

Fig. 1. Tracing showing inhibition of cardiac Ca^{2+} transients by TNFα. Indo-1 Ca^{2+} of an adult rat ventricular myocyte. Ca^{2+} transients were recorded from cardiac myocytes under electrical pacing conditions of 0.3 Hz. At approx 20 s, normal Tyrode's solution was removed and replaced with another dose of normal Tyrode's (sham) followed by the addition of TNFα (0.6 n*M*). Note that as expected TNFα caused a rapid, gradual reduction in transient amplitude, whereas the sham treatment had no effect. Washing the cells with normal Tyrode's restored the normal Ca^{2+} transient, which again was unaffected by the sham (normal Tyrode's) addition prior to the depression in transient amplitude associated with TNFα treatment.

1. Following overnight culture in DMEM/F12 in the presence or absence of cytokines, cells are rinsed thoroughly to remove cell debris and incubated in Ca^{2+}-free Tyrode's to minimize spontaneous myocyte contractility which would interfere with seal formation necessary for patch clamping (*see* **step 2**) and would hamper extraction of RNA.
2. Fire-polished, baked pipets with 5–10 microohm resistance are filled with sterile Ca^{2+}-free Tyrode's solution. Individual cells are patch-clamped in the whole cell configuration *(2,3)*.
3. After detection of whole-cell currents, negative pressure (suction to 9-mL mark on 10-mL syringe for 2–5 min) is applied to the pipet and a fraction of the cell's cytoplasmic contents are aspirated into the patch pipet tip which is then broken against the side of a 0.5-mL sterile microcentrifuge tube.
4. Approximately 2 μL of the pipet's contents are expelled and rapidly mixed into 7.5 μL of ice-cold first-strand cDNA synthesis mixture (*see* **Note 8**).
5. First-strand cDNA is synthesized by addition of 0.5 μL of AMV-RT (20 U), followed by incubation at 42°C for 1 h.
6. The reaction is terminated by incubating at 90°C for 10 min.

7. Aliquots of the cDNA template from first strand synthesis reaction are added to RT-PCR buffer and subjected to at least 30 cycles of PCR using a programmable thermal cycler (*see* **Note 9**).
8. Aliquots of PCR products are ethanol precipitated at –70°C, electrophoresed through a 2% agarose gel and visualized by ethidium bromide staining.
9. The remainder of the PCR product is ethanol precipitated and digested with specific restriction enzymes (depends on plasmid used). The required fragment is isolated following electrophoresis from a 2% LMP agarose gel and the product subcloned into a suitable vector and sequenced.

4. Notes

1. Important note: Be very careful and think ahead of all steps in the myocyte isolation procedure. **Steps 1–5** (**Subheading 3.1.**) should be completed in less than 1 min. Make sure that there are no bubbles introduced in the perfusion system.
2. Myocyte isolation (**Subheading 3.3., step 7**): This procedure works best with two people; one to hold the heart on the cannula and one to tie it down. As the perfusion starts, you should observe a clearing of the blood vessels.
3. Commercial collagenase preparations are occasionally contaminated with lipopolysaccharide that can adversely affect TFNα secretion and cardiomyocyte viability for subsequent experiments. It is recommended that batches of collagenase are routinely depyrogenated *(8)* to remove the lipopolysaccharide.
4. Myocyte isolation (**Subheading 3.1., step 8**): This step is crucial. Solution B contains collagenase which will digest the myocytes. Too little digestion will mean large aggregations of cells and thus poor isolation of individual myocytes, whereas overdigestion will result in poor cell viability. A good technique for determining when to stop the digestion is to look at the atrium of heart. Give a little tug after approx 25 min. If the atrium pulls easily, it means that the tissue has been digested enough and should be removed from the cannula.
5. Myocyte isolation (**Subheading 3.1., step 12**): The pipet tip should be cut to allow larger tissue chunks to be resuspended and facilitate digestion.
6. Myocyte isolation (**Subheading 3.3., step 13**): Looking at the cells is very important. You should make sure cells are not "clumped" together (underdigestion). Overdigestion is characterized by hypercontracted (round cells) and rod-shaped cells with frayed ends.
7. To minimize potential Ca^{2+} compartmentalization problems described by others, cells can be loaded for a brief (<15 min) periods of time. Even with these short loading times, it is possible that some mitochondrial Ca^{2+} can contribute to the baseline indo-1 fluorescence.
8. Two microliters of cell-conditioned Ca^{2+}-free Tyrode's solution should be used as a control to ensure that the RNA was not debris from damaged cells.
9. Specific primers for the desired gene product must be chosen, and corresponding denaturing, annealing and extending temperatures determined.

Acknowledgments

This work was supported by the American Heart Association National center (R.A.S.), the Muscular Dystrophy Association (R.A.S.), the National Institutes of Health (2SO6 GM 45765), the Diabetes Action Research and Education Foundation (K.A.K.), and the California Metabolic Research Foundation (K.A.K.). The authors gratefully acknowledge Chris Glembotski and Paul Paolini for advice, Pat McDonough for assistance with the Ca^{2+} transient experiments, and Russell Romeo and Betsy George for helpful discussions of the myocyte-isolation procedure.

References

1. Comstock, K. L., Krown, K. A., Page, M. T., Martin, D., Ho, P., Pedraza, M., Glembotski, C. C., Quintana, P. J. E., and Sabbadini, R. A. (1998) LPS-induced TNFα secretion and apoptosis in rat cardiomyocytes. *J. Mol. Cell. Cardiol.* **30,** 2761–2775.
2. Krown, K. A., Yasui, K., Brooker, M. J., Dubin, A. E., Nguyen, C., Harris, G. L., McDonough, P. M., Glembotski, C. C., Palade, P. T., and Sabbadini, R. A. (1995) TNFα receptor expression in cardiac myocytes: TNFα inhibition of L-type Ca^{2+} current and Ca^{2+} transients. *FEBS Lett.* **376,** 24–30.
3. McDonough, P. M., Yasui, K., Betto, R., Salviati, G., Glembotski, C. C., Palade, P. T., and Sabbadini, R. A. (1994) Control of cardiac Ca^{2+} levels: inhibitory actions of sphingosine on Ca^{2+} transients and L-channel conductance. *Circ. Res.* **75,** 981–989.
4. Smith, M. A. and O'Dowd, D. K. (1994) Cell-specific regulation of agrin RNA splicing in the chick ciliary ganglion. *Neuron* **12,** 795–804.
5. Krown, K. A., Page, M. T., Nguyen, C., Zechner, D., Gutierrez, V., Comstock, K. L., Glembotski, C. C., Quintana, P. J. E., and Sabbadini, R. A. (1996) TNFα-induced apoptosis in cardiac myocytes: involvement of the sphingolipid signaling cascade in cardiac cell death. *J. Clin. Invest.* **98,** 2854–2865.
6. Bailey, B. A., Dipla, K., Li, S., and Houser, S. R. (1997) Cellular basis of contractile derangements of hypertrophied feline ventricular myocytes. *J. Mol. Cell. Cardiol.* **29,** 1823–1835.
7. Zechner, D., Craig, R., Hanford, D., McDonough, P. M., Sabbadini, R. A., and Glembotski, C. C. (1998) MKK6 inhibits myocardial cell apoptosis via a p38 MAP kinase-dependent mechanism. *J. Biol. Chem.* **273,** 8232–8239.
8. Lew, W. Y. W., Lee, M., Yasuda, S., and Bayna, E. (1997) Depyrogenation of digestive enzymes reduces lipopolysaccharide tolerance in isolated cardiac myocytes. *J. Mol. Cell. Cardiol.* **29,** 1985–1990.

17

Culture of Primary Human Bronchial Epithelial Cells

Joanna Picot

1. Introduction

The human airway is normally kept sterile despite continual exposure to airborne pathogens. This is achieved by host defense mechanisms that exist to prevent adherence, colonization, and invasion of the airway epithelium. The airway is, however, a potential route by which bacteria may enter the bloodstream if host defense mechanisms can be overcome.

The airway epithelium is composed of several cell types and in section the cells appear to be arranged in layers. The more differentiated cells, e.g., ciliated cells and mucus-secreting cells, are at the surface, whereas the stem cells lie underneath. All cells, however, maintain their attachment to the basement membrane, and thus the airway epithelium is described as pseudostratified.

Whereas lung epithelial cell lines (e.g., A549, BEAS-2B) are available that grow and divide continuously and are easily maintained in culture, they may exhibit characteristics that are not present in the original primary cell type. Culture of primary cells is slightly more difficult. Primary cells have a finite life span but have the advantage that a proportion of the cells in primary culture retain the ability to differentiate. The characteristics of a culture of primary cells may therefore be a more accurate reflection of the characteristics exhibited by epithelial cells in vivo. Adherent cells such as epithelial cells are typically cultured in plastic tissue-culture flasks under a layer of suitable tissue-culture fluid. Lung epithelial cells will grow perfectly well in this type of culture; however, it does not resemble the growth conditions of this cell type in vivo. Yamaya et al. *(1)* have described the development of a culture system for human lung epithelial cells in which the apical surface of the cells is exposed to the air, whereas the basal surface is bathed in the tissue-culture medium. The

From: *Methods in Molecular Medicine, Vol. 36: Septic Shock*
Edited by: T. J. Evans © Humana Press Inc., Totowa, NJ

cells are seeded onto a permeable support that sits suspended in a well to which the tissue-culture medium is added so the cells then grow at the liquid air interface. This type of system is commercially available (e.g., Corning, Corning, NY; Costar, Cambridge, MA; Millipore, Bedford, MA). The point at which a fully confluent monolayer of cells develops tight junctions is established easily using this system because the media in the basal compartment will cease to seep through into the apical compartment. In addition Yamaya et al. report that epithelial cells grown with an air interface display improved cellular differentiation and have a multilayered appearance that more closely resembles the bronchial epithelium in vivo.

This chapter describes a method based on that described by Yamaya et al. for the isolation and culture of primary human bronchial epithelial cells (HBECs) which can be used as an in vitro system to study the interactions between pathogens and the airway epithelium.

The establishment and maintenance of a HBEC primary culture involves the following major procedures: dissociation of epithelial cells; seeding epithelial cells into culture vessels; cell passage; freezing cells; and replating frozen cells.

2. Materials

2.1. General Materials

The following materials were obtained from GibcoBRL (Life Sciences International, Paisley, UK) unless stated otherwise.

1. Hanks balanced salt solution (HBSS) without calcium and magnesium.
2. Dulbecco's modified Eagles medium (DMEM):F12 1:1 without glutamine.
3. Glutamine (200 m*M*).
4. Fungizone amphotericin B (250 µg/mL).
5. Penicillin-streptomycin (5000 U/mL to 5000 µg/mL).
6. Colistin methanesulfonate (sodium salt) (Sigma, Poole Dorset). Stock solution is 4 mg/mL stored at 4°C.
7. Fetal calf serum (FCS): 50-mL single-use aliquots stored at –20°C.
8. Ultroser G serum substitute.
9. Retinoic acid (Retinol Acetate, Sigma). Stock solution 330 n*M*, light-sensitive, stored at 4°C.

2.2. Materials for Dissociation of Epithelial Cells

All materials are available from Sigma.

1. Protease Type XIV.
2. Dithiothreitol (DTT).
3. Trypan Blue: stock solution 0.4%(w/v) in sterile water stored at room temperature.

2.3. Seeding Epithelial Cells into Culture Vessels and Replating Frozen Cells

All materials are available from Boehringer Mannheim (Indianapolis, IN).

1. Albumin (BSA): 100 mg/mL stored at 4°C.
2. Collagen S (Type 1): 3 mg/mL stored at 4°C.
3. Fibronectin: stock solution 1 mg/mL in sterile PBS stored at 4°C.

2.4. Cell Passage

1. Trypsin-EDTA (GibcoBRL): 5-mL single-use aliquots stored at 20°C.

2.5. Freezing Cells

1. Dimethyl sulphoxide (DMSO).
2. FCS (as above).

3. Methods

Perform all procedures in a laminar flow hood using sterile tissue-culture-grade plasticware and sterile dissection equipment. Prewarm media to 37°C before use unless stated otherwise.

3.1. Dissociation of Epithelial Cells

1. When the tissue sample is received, immediately remove it from the fluid in which it was transported. Rinse it four times in HBSS containing 5 m*M* DTT (make fresh as required) to remove mucus and associated debris from the tissue sample (**Note 1**).
2. Rinse the tissue twice in HBSS containing penicillin-streptomycin (50 U/mL to 50 μg/mL), colistin (4 μg/mL), and Fungizone amphotericin B (2.5 μg/mL) and keep it in this solution. For dissection transfer the tissue into a shallow dish. To ensure that the tissue does not dry out keep it fully submerged in HBSS containing antibiotics and antifungal as above. It may be necessary to pin the tissue down in a dissecting dish. The presence of antibacterial and antifungal agents at this early stage is important in combating the likelihood of later microbial growth.
3. Score the epithelium longitudinally into strips approx 3–5 mm wide using a scalpel and blade. The epithelium is stripped from the submucosa to minimize the presence of other cells, e.g., fibroblasts, in the preparation.
4. Pull epithelial strips away from the submucosa using forceps (**Note 2**).
5. As the strips are removed place them in HBSS containing antibacterial and antifungal agents.
6. Prepare a solution of protease (0.4 mg/mL in HBSS plus antibacterial and antifungal agents) and incubate the strips overnight at 4°C with continuous gentle agitation, e.g., on a rotating wheel. The protease solution must be made fresh each time it is required.

7. The following morning add FCS to a final concentration of 2.5% to inhibit the protease.
8. Dislodge the cells from the strips by vigorous agitation. Transfer the fluid into a centrifuge tube. Take care not to carry over any of the large denuded strips of tissue. To ensure that the majority of dissociated cells are collected add a further volume of HBSS to the strips, vigorously agitate again and transfer the fluid into the centrifuge tube.
9. Sediment the cells by centrifugation (200*g*, 10 min, 4°C).
10. Remove the supernatant carefully and then resuspend the pellet of cells in DMEM:F12 containing penicillin-streptomycin (50 U/mL to 50 μg/mL), colistin (4 μg/mL), Fungizone amphotericin B (2.5 μg/mL), and 5% FCS. Count the cells using a hemacytometer and estimate cell viability using trypan blue (**Note 3**).
11. Seed the cells into culture flasks as required (*see* **Subheading 3.2.**).

3.2. Seeding Epithelial Cells into Culture Vessels (see Note 4)

1. Calculate the total number of viable cells present. If cells have been purchased, information will be provided regarding the number of cells present and the likely plating efficiency.
2. Decide how many vessels can be seeded. BioWhittaker recommend seeding cells at 3500 cells per cm^2 of culture flask area and increasing this to 10,000 cells per cm^2 when using multiwell vessels that have a small culture area (less than 2 cm^2). Yamaya et al. recommend seeding onto membranes at 10^6 cells per cm^2. Such a high density of cells should saturate the available growth area and therefore there is no need for cell division to occur for a confluent monolayer to form. We have not always had sufficient cells to seed at this density but have had satisfactory results seeding at 10,000 cells per cm^2 and waiting for cells to grow and divide until a confluent layer is formed.
3. Coat the surface of plastic tissue-culture vessels to encourage cell attachment and growth before seeding the cells (*see* **Note 5**).
4. Dilute the cells as necessary with prewarmed media and then add the diluted cells to the culture vessels. Aim to add 1 mL of volume for every 5 cm^2 of growth area, i.e., add 5 ml to a 25 cm^2 flask.
5. After overnight incubation at 37°C change the culture medium replacing with DMEM:F12 containing penicillin-streptomycin (50 U/mL to 50 μg/mL), colistin (4 μg/mL), Fungizone amphotericin B (2.5 μg/mL), and 2% Ultroser G. Retinoic acid may also be added to the culture medium (**Note 6**). If microbial growth is apparent at this stage it is unlikely that this will be overcome. The inclusion of colistin in the growth media however has enabled successful growth of cells even from tissue sources likely to carry a heavy microbial population, e.g., from samples of cystic fibrosis bronchus. Feed the cells with the above medium every other day initially. The cells should have the cobblestone appearance typical of this type of epithelial cell (*see* **Note 7**). As cells approach confluence it will be necessary to feed them every day or to increase the volume of culture medium in the flask to 2 mL per 5 cm^2.

6. Passage the cells at approx 80% confluence. Cells grown in culture flasks must not be allowed to become fully confluent because this will induce terminal differentiation to a squamous phenotype. Terminally differentiated cells no longer divide. Cells grown on membranes with air on the apical side must become confluent in order for tight junctions to form. Cells are grown in this way for a particular experimental purpose and are not passaged.

3.3. Cell Passage

1. Rinse the cells twice with HBSS. Serum and some serum substitutes contain factors that inhibit trypsin; rinsing with HBSS will remove these inhibitory factors. If cells are grown in a serum-free medium there is no need to rinse them.
2. Add prewarmed trypsin-EDTA to the flask, ensuring that all the cells are coated with the solution and then remove almost all of it. Incubate the flask at 37°C for 3 min then view the cells using phase-contrast microscopy, monitoring every minute from then on until 90% of the cells have rounded up. Tap the flask sharply to detach the cells from the flask (*see* **Note 8**).
3. Once 90% of the cells have detached from the flask, add prewarmed media to the cells. Pipet the medium up and down to disperse cell clumps and wash loosely adherent cells away from the bottom of the flask. The presence of 2% Ultroser G in the medium will inhibit the trypsin. If a serum-free medium is being used a trypsin inhibitor, e.g., soybean trypsin inhibitor, must be added to the flask before the addition of the serum-free medium. The growth of the cells may be improved if, after the addition of media, the cells are pelleted by centrifugation, the supernatant (containing traces of trypsin) is removed and fresh medium added.
4. Count the cells and seed at an appropriate density into coated vessels as described in the previous method (*see* **Note 9**).

3.4. Freezing Cells

If more cells than are necessary for experimental purposes have been grown these can be frozen and stored in liquid nitrogen for later use.

1. Trypsinize the cells from the flask as described in **steps 1–3** of **Subheading 3.3.**
2. Transfer the cell suspension into a centrifuge tube and pellet the cells at 200*g*, 4°C for 10 min.
3. Keep the centrifuge tube on ice and remove the supernatant.
4. Resuspend the cell pellet in freezing medium composed of 90% v/v FCS plus 10% v/v DMSO.
5. Transfer the cells to a suitable cryovial and place the vial in a cell-freezing container (*see* **Note 10**).
6. Place the cell freezing container at –80°C for 24 h.
7. For long-term storage, transfer the vials to liquid nitrogen.

3.5. Replating Frozen Cells

To seed cells bought from a commercial supplier follow the instructions provided by the company.

1. Thaw the vial of frozen cells rapidly at 37°C. As soon as the last ice crystal disappears from the vial, place the vial on ice.
2. Transfer the cells to a centrifuge tube and carefully add 10 mL of ice-cold culture medium.
3. Pellet the cells by centrifugation at 200*g*, 4°C for 5 min.
4. Resuspend the cell pellet in culture medium. Count the viable cells and seed at the appropriate density in a prepared flask of a suitable size (according to the guidelines set out above in **Subheading 3.2.**, the method for seeding cells into culture vessels). Incubate at 37°C.
5. Replace the medium the following day and then continue to grow the cells as described above.

4. Notes

1. To have the best chance of success, tissue should be received as early as possible. We have exclusively used material removed from patients undergoing lung transplants. Postmortem tissue has been used by others who have obtained tissue within 12 h of death, but found that the greatest success was achieved when the postmortem interval was under 6 h *(2)*. The authors report that cultures could not be developed from more than 30% of the tissue obtained because of uncontrollable microbial growth. Tissue samples should be obtained in as sterile conditions as possible and stored/transported at 4°C in tissue-culture media containing antibiotics. If tissue is not available primary cells can be bought from specialist suppliers, e.g., BioWhittaker.
2. It is often easier to slip the scalpel blade under one end of each strip to tease it away from the submucosa. Forceps can then more easily grasp the end of the strip which should come away cleanly as there is a natural partition between the epithelial layer and the underlying tissue. If the strips break as they are being pulled away then the scalpel can be used again to create an end that the forceps can grasp.
3. Resuspend the cell pellet in a small volume, e.g. 0.5–1 mL for cells obtained from 2–3 cm^2 of tissue, otherwise the cell density may be too low for seeding and **Subheading 3.1., step 9** will have to be repeated. Mix a 10–15 μL volume of fully resuspended cells with an equal volume of trypan blue and count the cells using a hemocytometer. Viable cells will exclude the blue dye but nonviable cells will stain blue. We have observed that a good indication of the overall condition of the cells is the proportion of ciliated cells whose cilia are still actively beating.
4. The cells can be grown either in typical culture flasks, Petri dishes, and multiwell slides or they can be seeded onto a membrane to allow growth with an air inter-

face. The choice of culture vessel will in part depend on the type of experiment that will be undertaken but also on the number of cells isolated. It is more prudent to seed several small flasks than one large flask. In this way some flasks of cells may survive even if one is lost due to infection.

5. We use a mixture of collagen, fibronectin, and BSA (30 μg/mL collagen, 10 μg/mL fibronectin, and 100 μg/mL BSA in sterile PBS) to coat tissue-culture vessels. This coating mixture can be stored at 4°C and reused until it runs out or is no longer effective. Coated dishes can also be stored at 4°C. The coated dishes must be rinsed twice with HBSS to remove excess salt and should be allowed to come to room temperature (if they have been stored at 4°C) before use. The transwell membranes can be bought ready coated with collagen and need no further treatment.
6. A number of factors may influence the growth and potential differentiation of primary HBECs in culture. In particular retinoic acid (vitamin A) is known to inhibit terminal differentiation to the squamous cell type. Terminal differentiation occurs both in vivo and in vitro when vitamin A is lacking from diet or culture medium respectively. In addition, vitamin A is known to promote the expression of a mucosecretory phenotype (reviewed in **ref.** ***3***). If desired add vitamin A to the culture medium at a final concentration of 0.33 nM (equivalent to 0.1 μg/L).
7. In addition to the characteristic cobblestone morphology of the cells, their epithelial origins can also be verified by the presence of cytokeratin family of proteins. Cytokeratins are intermediate filament proteins expressed only in cells of epithelial origin. There are many different cytokeratins so it may be necessary to use a range of cytokeratin antibodies with differing specificities for immunochemistry ***(4)***. As the length of time in culture and passage number of the cells increases cytokeratin expression may be lost as the cells dedifferentiate. The cobblestone appearance of the cells will also be lost with continued culture. To ensure that the cells used for experimental purposes are as phenotypically similar to airway epithelial cells in vivo we have only used cells within the first two passages after isolation. In our hands these cell cultures, whether of cells isolated by us or whether obtained from a commercial supplier, have dedifferentiated by passage number 4 or 5.
8. Treatment of cells with trypsin is the most crucial step of the passage process. The solution of trypsin should be warmed to 37°C and used immediately so that the enzyme begins to act as soon as the solution meets the cells. If the trypsin solution is kept at 37°C for an extended period of time before being added to the cells it will have lost activity, the process will take longer and the yield of viable cells will be diminished. For a flask with a growing area of 75–80 cm^2 a volume of approx 1.5 mL of trypsin EDTA solution should remain in the flask during incubation. As the incubation proceeds the cells should start to round up, once the majority of the cells have rounded up tap the flask sharply and the cells will detach. Once 90% of the cells have detached add media to the flask as described in the method.

9. If the viability of the cells is poor following trypsinisation it is likely that the cells were exposed to the enzyme for too long. If the yield of cells is low this may be caused by incomplete trypsinisation of cells—check the flask after trypsin treatment to see whether many cells have remained attached to the base.
10. The contents of the cryovials need to freeze slowly and steadily and so should be placed in an insulated freezing container. Special containers can be bought which consist of two compartments (Sigma, Poole, UK). The lower compartment is filled with alcohol, whereas the vials sit in depressions in the upper compartment, an arrangement that causes the cells to cool and freeze slowly. A box made from an insulating material, e.g., polystyrene, provides a suitable (and cheaper) alternative to a bought freezing container.

References

1. Yamaya, M., Finkbeiner, W. E., Chun, S. Y., and Widdicombe, J. H. (1992) Differentiated structure and function of cultures from human tracheal epithelium. *Am. J. Physiol.* **262,** L713–L724.
2. Elliget, K. A. and Lechner J. F. (1992) Normal human bronchial epithelial cell cultures, in *Culture of Epithelial Cells* (Freshney, R. I., ed.) Wiley-Liss, New York, pp. 181–196.
3. Jetten, A. M. (1991) Growth and differentiation factors in tracheobronchial epithelium. *Am. J. Physiol.* **260,** L361–L373.
4. Mackenzie, I. C. and Gao, Z. (1996) Identification and localisation of cytokeratins, in *Epithelial Cell Culture: A Practical Approach* (Shaw, A. J., eds.), IRL Press, New York, pp. 17–36.

18

Primary Culture of Human Proximal Renal Tubular Epithelial Cells

Paul A. Glynne

1. Introduction

Sepsis and septic shock are major causes of acute renal failure (ARF). Although hemodynamic factors play a significant role in the pathogenesis of ARF during sepsis, it is now clear that nonhemodynamic factors are also extremely important. The predominant site of tissue injury in sepsis-induced ARF occurs within the proximal renal tubule. In vivo studies of the specific cellular mechanisms underlying renal injury are limited by the marked heterogeneity of the nephron. Establishing primary cultures of human proximal renal tubular epithelial cells (PTEC) provides a well-characterized in vitro model, phenotypically representative of PTEC in vivo. This in vitro system allows for investigation of the cellular mechanisms underlying proximal tubular injury during sepsis, in isolation without additional complicating cardiovascular and neuroendocrine factors.

1.1. Renal Tubular Epithelial Cells in Culture

The renal tubule contains a number of different segments and approx 15 epithelial cell types, the majority of which have been cultured successfully. Importantly, each epithelial cell type demonstrates specific morphological, biochemical, and functional properties that allows for their precise characterization.

Common to all epithelia, PTEC are polarized, with a basolateral plasma membrane attached to the basement membrane or culture growth matrix, and an apical/luminal membrane supporting microvilli. Tight junctions (zonula occludens) are located at the junction of the apical and basolateral membranes. The tight junctions produce trans-epithelial resistance and limit paracellular

From: *Methods in Molecular Medicine, Vol. 36: Septic Shock*
Edited by: T. J. Evans © Humana Press Inc., Totowa, NJ

ion and fluid transport. There is a highly organized cytoskeleton with cytokeratin intermediate filaments in the cell body.

1.2. Proximal Renal Tubular Epithelial Cells in Culture

The most important function of the proximal tubule is the iso-osmotic reabsorption of 7/8 of the glomerular filtrate. It therefore has all the characteristics of an actively transporting epithelium with an increased apical and basolateral surface area, and numerous mitochondria. The cells are irregular in shape and interdigitate freely, a feature that can be observed under scanning electron microscopy. The luminal surface area is increased by numerous microvilli forming the brush border, which expresses γ-glutamyl transpeptidase, alkaline phosphatase, glucose-6-phosphatase, and leucine aminopeptidase. In primary culture, PTEC do not display significant brush-border formation unless the cells are grown on semipermeable supports that promote monolayer polarization, tight-junction formation (measured by the development of transepithelial resistance) and greater cellular differentiation. Therefore brush-border enzyme activity can be difficult to detect and may indeed decrease significantly during monolayer growth *(1)*, if the cultures are not grown in optimal conditions.

PTEC express various transporters (apical and basolateral sodium-hydrogen exchangers driven by Na^+/K^+-ATPase) and cotransporters (including Na^+-glucose, Na^+-amino acid, and Na^+-phosphate) *(2)*. PTEC Na^+-dependent phosphate transport is inhibited by parathormone (PTH), which stimulates cyclic adenosine monophosphate (cAMP) production. cAMP stimulation by PTH and not arginine-vasopressin (AVP) or calcitonin is a specific proximal tubular hormonal response and is extremely important in the characterization of PTEC in culture.

1.3. Principles of PTEC Culture

The optimal method for PTEC culture should be simple to perform and yield a large homogeneous cell population. A variety of techniques have been established for the primary culture of human PTEC. Cultures from explants and cell suspensions often combined with additional purification steps yield high cell numbers but may not be completely homogeneous. Techniques such as nephron microdissection and immunodissection using cell-type specific antibodies yield very pure cell populations but relatively low numbers of cells.

In this chapter, I will detail the methods used within our laboratory to consistently obtain large numbers of human renal tubular epithelial cells with a proximal tubular phenotype. The techniques described have been adapted from previously published methods *(1,3,4)*. The technique relies on the fact that the bulk of proximal tubular epithelia are especially concentrated in the extreme outer renal cortex. Enzymatic digestion of the outer renal cortex followed by sequential sieving, to remove tubular fragments and glomeruli, leaves material that yields outgrowth of renal epithelial cells with a proximal phenotype. The culture of these cells in a hormonally defined media, on a collagen/fetal calf serum matrix encourages PTEC growth and repeated subculture. Using serum-free media inhibits the overgrowth of fibroblasts.

The methods and protocols can be usefully subdivided into the following categories:

1. Isolation of PRTEC from human renal cortex.
2. Maintenance of cells in culture and cell passage.
3. Morphological, biochemical and functional characterization of cells.
4. Cell passage.
5. Cell cryopreservation, thawing and replating.

2. Materials

The following materials were obtained from GibcoBRL (Life Sciences International, Paisley, UK) unless stated otherwise. Store at either room temperature (RT), 4°C, –20°C, or –80°C, as indicated.

2.1. Materials for Cell Isolation from Human Renal Cortex

1. Type II collagenase from *Clostridium histolyticum* 1 mg/mL (Sigma) –20°C.
2. Trypsin-EDTA: 0.05% trypsin, 0.02% EDTA; 5-mL aliquots stored at –20°C.
3. Cell strainers, sieve sizes 100 and 40 um (Becton Dickinson, Oxford, UK).

2.2. Media Supplements

1. Insulin-transferrin-selenium supplement (ITS) (Boehringer Mannheim, Lewes, East Sussex, UK) (stock 100×) at 4°C.
2. Hydrocortisone (Sigma) (stock 360 µg/mL in ethanol) at –20°C.
3. Triiodothyronine (Sigma) (stock 0.4 µg/mL in water) store in aliquots at –20°C.
4. Human epidermal growth factor (EGF) (Peprotech, London, UK) (stock 100 µg/mL) store in aliquots at –80°C for a maximum of 6 mo.
5. Glutamine (200 m*M*) at –20°C.
6. Penicillin-Streptomycin (5000 U/ml to 5000 µg/mL) at –20°C.

2.3. Media

1. Hanks balanced salt solution (HBSS) without calcium and magnesium at RT.
2. Dulbeco's modified Eagles medium (DMEM)/ Ham's F-12 (1:1 mixture) at 4°C.
3. Serum-free media: DMEM/Ham's F-12 (1:1 mixture) containing glutamine (2 m*M*), penicillin-streptomycin (50 U/mL to 50 µg/mL), insulin (10 µg/mL),

transferrin (5 µg/mL), selenium (5 ng/mL), hydrocortisone (36 ng/mL), triodothyronine (4 pg/mL), and EGF (10 ng/mL).

2.4. Matrix Substrate

1. Collagen S, from calf skin (3 mg/mL) (Boehringer Mannheim, Lewes, East Sussex, UK) at 4°C.
2. Fetal calf serum (FCS) at 4°C.
3. Phosphate-buffered saline (PBS) at RT.

2.5. Cell Passage

1. Trypsin-EDTA 5-mL aliquots stored at –20°C.
2. Soybean trypsin inhibitor (Boehringer Mannheim) stored as powder at –20°C, reconstituted in filter-sterilized PBS (0.1%), and stored at 4°C (maximum 2 wk).

2.6. Cell Cryopreservation

1. 10% dimethyl sulphoxide (DMSO) in serum-free media at 4°C.

2.7. Cell Characterization

2.7.1. Electron Microscopy (EM)

1. 3% EM-grade glutaraldehyde, 0.1 *M* sodium cacodylate buffer at pH 7.2.
2. 1% osmium tetroxide (Agar Scientific) at 4°C.

2.7.2. Immunocytochemistry

1. Monoclonal anti-pan cytokeratin antibody (Sigma, clone C-11) at –80°C.

2.7.3. Hormonal Response

1. Human parathyroid hormone (PTH) at –80°C.
2. Arginine-vasopressin (AVP) at –80°C.
3. Calcitonin and 3-Isobutyl-1-methylxanthine (IBMX) at –80°C.
4. cAMP enzyme immunoassay system (Amersham) at 4°C.

3. Methods

Perform all procedures in a laminar flowhood using sterile tissue-culture grade plastic ware and sterile dissection equipment. Prewarm media and HBSS to 37°C prior to use unless stated otherwise.

3.1. Preparation of Matrix Substrate

1. Coat tissue-culture flasks (use 25-cm^2 flasks for seeding cells directly from enzymatically digested cortex) with collagen S (0.25 mg/mL in filter-sterilized PBS) and allow to dry in the flow hood.
2. Add FCS to the flasks so that the entire surface is coated and then incubate at 37°C for 6 h.

3. Transfer flasks from the incubator to the flow hood and remove the FCS. Wash the flask surface three times with HBSS. The flasks are now ready for use although can be stored at 4°C for up to 2 wk.

3.2. Isolation of Cells from Human Renal Cortex

Human renal cortex is obtained from nephrectomy specimens performed for resection of renal tumors. A core of normal cortex from the disease-free pole, surplus to clinical requirement, is removed.

1. Collect the tissue immediately into HBSS containing penicillin-streptomycin (50 U/mL to 50 mg/mL) and place on ice during transfer to the laboratory (*see* **Note 1**).
2. When the kidney arrives at the tissue-culture facility, place the specimen in a shallow Petri dish in the flowhood and add HBSS containing penicillin-streptomycin now warmed to 37°C.
3. Using sterile dissecting equipment strip the fibrous capsule away from the cortex. The capsule should peel off very easily using forceps only.
4. Dissect away the very outer cortex from the inner cortex and medulla.
5. Cut the tissue sample with scissors and forceps into pieces approx 1 mm^3 (*see* **Note 2**).
6. Transfer the tissue fragments to a sterile 50-mL centrifuge tube and wash with HBSS three times. Washing involves 5-min spins at 200*g*, pouring off supernatant, and resuspending in HBSS.
7. Add collagenase 1 mg/mL to the 50-mL tube so that all the fragments are covered by the solution. Incubate at 37°C for 60 min with intermittent agitation of the tube.
8. Transfer the tube to the flow hood and add HBSS to make a final volume of 30 mL.
9. Force the tissue fragments within the 30-mL volume of HBSS through a 100-μm sieve into a 50-mL centrifuge tube. Then pass sieved material through a 40-μm sieve into another 50-mL centrifuge tube. This removes tubular fragments and glomeruli respectively.
10. Sediment the sieved cells by centrifugation (200*g*, 5 min at RT).
11. Pour off supernatants and resuspend the cell pellet in serum-free, hormonally defined media. Cells are then counted using a hemacytometer. Transfer to 25-cm^2 tissue-culture flask at a cell density of 20,000 cells per cm^2. Incubate at 37°C in a 5% CO_2: 95% air incubator.
12. The remaining tissue fragments, excluded by sieving, can then be further digested by incubating with trypsin-EDTA for a further 45 min at 37°C. Trypsin activity is terminated after this time by adding an equal volume of 0.1% soybean trypsin inhibitor. Following this, proceed through **steps 9–11** and plate cells out at the recommended seeding density (*see* **Note 3**).

3.3. Maintenance of Cells in Culture

Using the isolation protocol above, cells undergoing first passage should form confluent monolayers 5–10 d from initial seeding into flasks. The media should be changed at 24 h after which time small islands of attached cells are visible under phase-contrast microscopy. Thereafter, the media should be exchanged every 48 h until the monolayers are 50–60% confluent. Once over 50% confluent, the rate of cell proliferation increases significantly necessitating media changes every 24 h until confluence is reached. Once confluent, the cells can be trypsinised and passaged, as described below, further subcultured or seeded into a suitable vessel for the purposes of experimentation. Cells beyond the second passage reach confluence in 2–5 d (*see* **Note 4**).

3.4. Characterization of Human Proximal Renal Tubular Epithelial Cells

It is imperative that cell phenotype is established prior to the use of these cells for the purposes of investigation. The following provides an overview of the methods used in our laboratory. It is not intended to be exhaustive, and the reader is referred on to the relevant reference source where appropriate.

3.4.1. Morphology and Ultrastructure

Under phase-contrast microscopy, confluent primary cultures demonstrate a typical cobblestone appearance. After 5 d postconfluence, areas develop within the monolayer where the cells heap up, forming hemicysts or domes; this indicates active fluid transport with accumulation of media between the basal aspect of the monolayer and the culture flask FCS/collagen matrix. The ultrastructure of the cells can be demonstrated on scanning and transmission electron microscopy. PTEC have a polarized morphology with numerous apical microvilli; cells contain numerous mitochondria and abundant rough endoplasmic reticulum ***(4–6)*** (*see* **Note 5**).

3.4.2. Biochemical Characteristics

Well-differentiated PTEC cultures (i.e., monolayers grown on porous inserts) express a number of brush-border enzymes (*see* **Subheading 1.2.**) for which there are a number of assay methods ***(1,3–7)***.

1. Immunochemical characterization: PTEC demonstrate positive immuno-staining for cytokeratin (*see* **Note 6**).
2. Hormonal response: Intracellular concentrations of cAMP are markedly and significantly increased in PTH-treated cells compared with controls or cells incubated with AVP or calcitonin (*see* **Note 7**).
3. Functional characteristics: PTEC monolayers grown on porous inserts achieve

maximal transepithelial resistance (approx 100 ohm.cm^2) 5 d postconfluence. They demonstrate phlorizin-inhibitable, Na^+-dependent uptake of a-methyl-D-glucopyranoside, a nonmetabolizable substrate for the Na^+-dependent hexose transport system *(2)*.

3.5. Cell Passage

Using these methods, PTEC can be repeatedly subcultured at a 1:3 ratio for up to eight passages with the cells maintaining their proximal tubular phenotype. Cell passage is performed as follows:

1. Rinse the cells twice with HBSS without calcium and magnesium. This removes all potential factors which can reduce trypsin-EDTA activity (e.g., calcium/magnesium-chelation of EDTA).
2. Add prewarmed trypsin-EDTA to the flask ensuring that all cells are coated with the solution. Remove the majority of the trypsin and incubate the cells at 37°C for 4 min. Observe the cells intermittently under phase-contrast microscopy to ensure adequate cell detachment.
3. After 4 min, cell detachment is further facilitated by striking the culture flask against the palm of the hand.
4. Terminate trypsin action by adding an equal volume of 0.1% soybean trypsin inhibitor.
5. Resuspend the detached cells in media and distribute to culture flasks at a 1:3 subculture ratio.

3.6. Cell Cryopreservation, Thawing, and Replating

3.6.1. Cryopreservation

1. Trypsinize confluent monolayers as described above.
2. Transfer the cell suspension into a centrifuge tube and pellet the cells at 200*g*, 4°C for 5 min.
3. Transfer the tube containig the cell pellet to ice.
4. Resuspend the cell pellet in serum-free growth medium containing 10% DMSO.
5. Transfer the cells to a cryovial and place within a cell freezing container in a –80°C freezer for 24 h (*see* **Note 8**).
6. Transfer to liquid nitrogen for long-term storage (*see* **Note 9**).

3.6.2. Thawing and Replating Cells

1. Thaw the vial of frozen cells rapidly at 37°C. Transfer the vial to ice immediately on thawing.
2. Transfer the cells to a centrifuge tube and add 10-mL of ice-cold culture medium.
3. Pellet the cells by centrifugation at 200*g*, 4°C for 5 min.
4. Resuspend the cell pellet in warmed culture medium and seed at the appropriate cell density into a preprepared tissue-culture flask of a suitable size (*see* **Note 10**).

4. Notes

1. Collect the specimen as soon after resection as possible as there is a clear decline in the amount of successfully cultured material the longer the tissue remains ex vivo (unless the kidney is in perfusate, as would be the case for a renal allograft).
2. Use blunt dissection wherever possible to minimize tissue damage. Be absolutely sure to remove the capsule in its entirety. Ensure that you use only the extreme outer cortex or it is very likely that the cell culture will be very heterogeneous. When mincing the specimen, try to cut the pieces as small as possible in order to minimize tissue waste (especially when sieving the tissue) and maximize collagenase digestion.
3. It is generally possible to grow up a confluent T25-cm^2 flask from even the smallest cell pellet (even though this may represent suboptimal cell density) so it is always worth attempting to seed whatever material remains.
4. The recommended cell density for seeding into any form of culture vessel for the purposes of experimentation (e.g., 24-well plate) is 10^5 cells/mL.
5. For SEM, cells are grown to confluence on collagen coated PTFE transwell inserts (Corning Costar) (6.5-mm diameter, 0.4-µm pore size) as assessed morphologically and following the development of sustained trans-epithelial resistance (measured with the Epithelial Voltohmmeter [World Precision Instruments]). Cells are prefixed in 3% EM-grade glutaraldehyde in 0.1 *M* cacodylate buffer pH 7.2, for 60 min; postfixed in 1% osmium tetroxide; dehydrated in a graded series of alcohols and then critical point dried from CO_2. For TEM, cells grown to confluence in 5-cm Petri dishes are scraped up into PBS, pelleted gently and then subjected to the pre- and postfixation described above. Cells are then embedded in araldite resin and sectioned prior to EM.
6. Immunostaining is performed on 1% paraformaldehyde-fixed confluent monolayers grown on chamber slides. Cells are permeabilized in 0.2% Triton, blocked in 2% horse serum, and then incubated with primary antibody (monoclonal anti-cytokeratin). Following PBS washes, the cells are incubated with a fluorescein-conjugated anti-mouse secondary antibody and then viewed under standard fluorescence microcopy. Mouse IgG1 is used as a control.

   URO-10 is another cytokeratin marker specific for the proximal tubule. PTEC will have negative immunostaining for vimentin and the distal tubular marker Tamm-Horsfall protein is negative. These markers serve as useful negative controls.
7. To measure the hormonal response, we use a method modified from that described by Schmouder et al. ***(8)***. In brief, PTEC are grown to confluence in four 9-cm Petri dishes. Cells are then stimulated for 15 min by either $10^{-7}M$ PTH, $10^{-6}M$ AVP or $10^{-6}M$ calcitonin, with one dish used as a control. Cells are then scraped into 0.5-mL ice-cold buffer (150 m*M* KCl, 5 m*M* K_2PO_4, 2 m*M* EDTA, and 0.5 m*M* IBMX (an inhibitor of cAMP phosphodiesterases), sonicated and boiled for 5 min. cAMP concentrations are measured using an enzyme immunoassay system (*see* **Subheading 2.**) ***(8)***.

8. To minimize loss of cell viability during the cryopreservation process, the cells are initially cooled slowly at a rate of –1°C/min in a (NALGENE) 1°C freezing container with isopropanol placed in a –80°C freezer.
9. Cells can be stored in liquid nitrogen and be successfully thawed and replated for up to 10 yr.
10. As a general rule, the cells can be successfully replated into the same size tissue-culture flask from which they were initially trypsinized and cryopreserved.

References

1. Triffilis A. L., Regec, A. L., and Trump, B. F. (1984) Isolation, culture and characterization of human renal tubular epithelial cells. *J. Urol.* **133,** 324–329.
2. Johnson D. W., Brew, B. K., Poronnik, P., Cook, D. I., Gyory, A. Z., Field, M. J., and Pollock, C. A. (1997) Transport characteristics of human proximal tubule cells in primary culture. *Nephrology* **3,** 183–194.
3. Detrisac, C. J., Sens, M. A., Garvin, J., Spicer, S., and Sens, D. A. (1984) Tissue culture of human kidney epithelial cells of proximal tubule origin. *Kidney Int.* **25,** 383–390.
4. Courjault-Gautier F., Chevalier, J., Abbou, C. C., Chopin, D. K., and Toutain, H. J. (1995) Consecutive use of hormonally defined serum-free media to establish highly differentiated human renal proximal tubule cells in primary culture. *J. Am. Soc. Nephrol.* **5,** 1949–1963.
5. Kempson, S. A., McAteer, J. A., Al-Mahrouq, H. A., Dousa, T. P., Dougherty, G. S., and Evan, A. P. (1988) Proximal tubule characteristics of cultured human renal cortex epithelium. *J. Lab. Clin. Med.* **113,** 285–295.
6. Wilson P. D., Dillingham, M. A., Breckon, R., and Anderson, R. J. (1985) Defined human renal tubular epithelia in culture: growth, characterization, and hormonal response. *Am. J. Physiol.* **248,** F436–F433.
7. Van der Biest I., Nouwen, E. J., Van Dromme, S. A., and De Broe, M. E. (1994) Characterization of pure proximal and heterogeneous distal human tubular cells in culture. *Kidney Int.* **45,** 85–94.
8. Schmouder, R. L., Strieter, R. M., Wiggins, R. C., Chensue, S. W., and Kunkel, S. L. (1992) In vitro and in vivo interleukin-8 production in human renal cortical epithelia. *Kidney Int.* **41,** 191–198.

Index

From: *Methods in Molecular Medicine*, Vol. 36: *Septic Shock*
Edited by: T. J. Evans © Humana Press Inc., Totowa, NJ

FSC
www.fsc.org
MIX
Papier | Fördert gute Waldnutzung
FSC® C083411

Kolibri 360 GmbH
Ferdinand-Jühlke-Straße 7
99095 Erfurt, Deutschland
produktsicherheit@kolibri360.de